PLCs for Beginners

An introductory guide to building robust PLC programs
with the Structured Text language

M. T. White

‹packt›

PLCs for Beginners

Group Product Manager: Preet Ahuja

Publishing Product Manager: Suwarna Rajput

Book Project Manager: Uma Devi

Senior Editor: Isha Singh

Technical Editor: Yash Bhanushali

Copy Editor: Safis Editing

Proofreader: Isha Singh

Indexer: Subalakshmi Govindhan

Production Designer: Vijay Kamble

DevRel Marketing Coordinators: Linda Pearlson and Rohan Dobhal

First published: May 2024

Production reference: 1100524

Published by Packt Publishing Ltd.
Grosvenor House
11 St Paul's Square
Birmingham
B3 1RB, UK

ISBN 978-1-80323-093-1

www.packtpub.com

To mom and dad.

Contributors

About the author

M. T. White has been programming since the age of 12. His fascination with robotics flourished when he was a child programming microcontrollers. He holds an undergraduate degree in mathematics, a master's degree in software engineering, and an MBA in IT management. He is currently working as a software developer for a major US defense contractor and is an adjunct CIS instructor, where he teaches Python, C, and an array of other courses. His background mostly stems from the automation industry where he programmed PLCs and HMIs for many different types of applications. He has programmed many different brands of PLCs over the years and has developed HMIs using many different tools. Other technologies that he is fluent in include Linux, Ansible, Docker, AWS, C#, Java, and Python. Be sure to check out his channel AlchemicalComputing on YouTube.

About the reviewers

Keith Lyding is an electrical engineer for a manufacturing company in Columbus, Ohio. He has over 15 years of experience in the electrical field, as well as more than 9 years of experience in automation. He graduated from Thomas Edison State University in 2019. He served in the US Navy for six years and has also worked for Nucor Steel. He currently works for Sonoco Products Company, where he works primarily with Allen Bradley PLCs, Inductive Automation's Ignition platform, EXOR and Panelview HMIs, and many other platforms. He enjoys troubleshooting, as well as automating complex operations. In his spare time, he loves to serve in his church, coach his son's baseball team, and spend time with his family.

I extend my deepest gratitude to my wife, Katie, for her unending support. I am grateful to my colleagues and mentors for their guidance throughout my journey: Kyle Ahrendt, William Carleton, and K. Andy Steinacker. I am also appreciative of the men in my life for continuing to sharpen, push, and encourage me to become more like Jesus: Paul Cassidy, Brian Babin, Ken Lyding, K. Andy Steinacker, Jon Krull, David Bout, Isaac Dye, and many, many others.

Ninad Deshpande is an author, storyteller, international speaker, and technology evangelist known in the industrial automation fraternity. He has over 15 years of extensive hands-on experience in varied fields of the automation industry, such as application development, testing, R&D, marketing, corporate communication, and global product management. He is the co-author of the Packt book *The Art of Manufacturing*. Today, as a co-founder and director of Passion Minds Private Limited, he helps organizations across the globe in various industries with services focusing primarily on technologically and strategically driven content generation.

I thank my family and friends for teaching me life lessons and consistently inspiring me. I admire my mother's and my wife's fighting spirit; their constant support and motivation enable me to achieve my dreams.

Table of Contents

2

3

4

Unleashing Computer Memory   45

5

Designing Programs – Unleashing Pseudocode and Flowcharts   57

6

Part 2: Introduction to Structured Text Programming

7

8

Exploring Variables and Tags   115

9

Performing Calculations in Structured Text   131

15

Secure PLC Programming – Stopping Cyberthreats   251

16

Troubleshooting PLCs – Fixing Issues   273

17

Leveraging Artificial Intelligence (AI)   293

18

The Final Project – Programming a Simulated Robot   307

Preface

Until recently, automation programming has been, for the most part, unchanging. However, with the recent boom in computing power, that is rapidly changing. New technologies are being introduced at a rapid pace, and these are drastically altering the automation landscape. These new, disruptive technologies are rendering the days of only programming in Ladder Logic a relic of the past. In today's automation landscape, to get the most out of a PLC, one must use Structured Text. This book is an in-depth look at writing very robust and well-written programs in Structured Text while also providing a general education for programming logic and design as well as other core tenets that will be required to future-proof projects.

Who this book is for

This book is for anyone who is interested in learning Structured Text programming. It is designed for beginners who have never programmed before and for those who wish to transition from Ladder Logic to Structured Text.

What this book covers

Chapter 1, *Computer Science Versus Automation Programming*, introduces you to computer science and contrasts it with automation programming. This chapter will explore the various types of controllers, emerging technologies, and more.

Chapter 2, *PLC Components – Integrating PLCs with Other Modules*, focuses on introducing the various components that make up a traditional PLC. This chapter will introduce you to analog and digital principles as well as all the needed components that a PLC will need to operate.

Chapter 3, *The Basics of Programming*, lays the foundation for programming. This chapter will introduce you to what programs are, how they work, and much more.

Chapter 4, *Unleashing Computer Memory*, lays the foundation for more advanced chapters by introducing you to the basics of memory. Topics explored will include what memory is, how memory works, and common storage devices.

Chapter 5, *Designing Programs – Unleashing Pseudocode and Flowcharts*, teaches you how to create a design for a program. Concepts explored will be designing a program in pseudocode and with a flowchart.

Chapter 6, *Boolean Algebra*, covers the basics of Boolean algebra. The core principles will be to learn how to compute logical equations, understand logical operators, and create truth tables.

Chapter 7, Unlocking the Power of ST, explores what Structured Text truly is and why it is important. The key takeaways from this chapter are understanding why Structured Text is important, why it should be used, and how to set up the programming environment.

Chapter 8, Exploring Variables and Tags, expands on and applies the material presented in *Chapter 4* to implement what are called variables or tags. This chapter will cover concepts such as data types, naming conventions, and much more.

Chapter 9, Performing Calculations in Structured Text, covers one of the most pivotal skills any PLC programmer can have: programming mathematical calculations. Topics will include how to program math equations and common math functions.

Chapter 10, Unleashing Built-In Function Blocks, explores the built-in function blocks. The main takeaway will be for you to understand what a built-in function block is and how to use common function blocks such as timers and counters.

Chapter 11, Unlocking the Power of Flow Control, introduces flow control with conditionals. This chapter will explore how the flow of a program can be altered and basic intelligence can be introduced to a program.

Chapter 12, Unlocking Advanced Control Statements, expands on concepts that were explored in the previous chapter, and examines topics such as embedded conditional statements, complex logical expressions, and much more.

Chapter 13, Implementing Tight Loops, provides an in-depth exploration of loops. This chapter will explore various types of loops in Structured Text as well as their applications.

Chapter 14, Sorting with Loops, introduces you to the basics of sorting algorithms. The key takeaway from this chapter is to introduce you to concepts such as algorithms, Big O notation, the basics of arrays, common sorting algorithms, and more.

Chapter 15, Secure PLC Programming – Stopping Cyberthreats, provides an overview of the cybersecurity landscape in relation to PLC-based systems. This chapter will present a lot of theoretical knowledge that can be applied to the design of PLC-based systems, networks, and more.

Chapter 16, Troubleshooting PLCs – Fixing Issues, provides the necessary steps to troubleshoot a malfunctioning PLC-based machine. Topics will include common issues, necessary tools, IT diagnostics, and more.

Chapter 17, Leveraging Artificial Intelligence (AI), explores how generative AI (ChatGPT) can be used to help automatically write software. This chapter will explore what generative AI is, how to use it, reasonable expectations, and how to write prompts.

Chapter 18, The Final Project – Programming a Simulated Robot, draws on material explored throughout the book. This chapter will focus on programming a theoretical robot that sorts parts and sends lots down the proper production line.

To get the most out of this book

This book assumes no prior knowledge of PLC programming or programming in general. To get the most out of this book, only a basic understanding of mathematics is required. Ideally, you should be familiar with basic algebra and maybe trigonometry.

Software/hardware covered in the book	Operating system requirements
CODESYS	Windows
ChatGPT	N/A

If you are using the digital version of this book, we advise you to type the code yourself or access the code from the book's GitHub repository (a link is available in the next section). Doing so will help you avoid any potential errors related to the copying and pasting of code.

Download the example code files

You can download the example code files for this book from GitHub at `https://github.com/PacktPublishing/PLCs-for-Beginners`. If there's an update to the code, it will be updated in the GitHub repository.

We also have other code bundles from our rich catalog of books and videos available at `https://github.com/PacktPublishing/`. Check them out!

Conventions used

There are a number of text conventions used throughout this book.

`Code in text`: Indicates code words in text, database table names, folder names, filenames, file extensions, pathnames, dummy URLs, user input, and Twitter handles. Here is an example: "We had a `CASE` conditional with an `IF` statement inside it ."

A block of code is set as follows:

```
PROGRAM PLC_PRG
VAR
currentHopperWeight     : REAL := 250;
bag1Weight              : REAL;
bag2Weight              : REAL;
END_VAR
```

When we wish to draw your attention to a particular part of a code block, the relevant lines or items are set in bold:

```
PROGRAM PLC_PRG
VAR
    password    : STRING(255) := 'password';
    length    : UINT;
    acceptPass : bool;
END_VAR
```

Bold: Indicates a new term, an important word, or words that you see onscreen. For instance: "**Air gapped systems** are simply systems that are not connected to the internet."

> **Tips or important notes**
> Appear like this.

Get in touch

Feedback from our readers is always welcome.

General feedback: If you have questions about any aspect of this book, email us at customercare@packtpub.com and mention the book title in the subject of your message.

Errata: Although we have taken every care to ensure the accuracy of our content, mistakes do happen. If you have found a mistake in this book, we would be grateful if you would report this to us. Please visit www.packtpub.com/support/errata and fill in the form.

Piracy: If you come across any illegal copies of our works in any form on the internet, we would be grateful if you would provide us with the location address or website name. Please contact us at copyright@packt.com with a link to the material.

If you are interested in becoming an author: If there is a topic that you have expertise in and you are interested in either writing or contributing to a book, please visit authors.packtpub.com.

Share your thoughts

Once you've read *PLCs for Beginners*, we'd love to hear your thoughts! Scan the QR code below to go straight to the Amazon review page for this book and share your feedback.

`https://packt.link/r/1803230932`

Your review is important to us and the tech community and will help us make sure we're delivering excellent quality content.

Download a free PDF copy of this book

Thanks for purchasing this book!

Do you like to read on the go but are unable to carry your print books everywhere?

Is your eBook purchase not compatible with the device of your choice?

Don't worry, now with every Packt book you get a DRM-free PDF version of that book at no cost.

Read anywhere, any place, on any device. Search, copy, and paste code from your favorite technical books directly into your application.

The perks don't stop there, you can get exclusive access to discounts, newsletters, and great free content in your inbox daily

Follow these simple steps to get the benefits:

1. Scan the QR code or visit the link below

https://packt.link/free-ebook/9781803230931

2. Submit your proof of purchase
3. That's it! We'll send your free PDF and other benefits to your email directly

Part 1: Basics of Computer Science for PLC Programmers

Programming is much more than just writing code. There is a lot of theoretical knowledge that goes into crafting a well-written program, and this part will lay the foundation for just that. This part will provide all the theoretical knowledge needed to understand the rest of the book and will cover the basics of PLCs, including PLC hardware, memory, logic, and design, how programs work under the hood, Boolean algebra, and more.

This part has the following chapters:

- *Chapter 1, Computer Science Versus Automation Programming*
- *Chapter 2, PLC Components – Integrating PLCs with Other Modules*
- *Chapter 3, The Basics of Programming*
- *Chapter 4, Unleashing Computer Memory*
- *Chapter 5, Designing Programs – Unleashing Pseudocode and Flowcharts*
- *Chapter 6, Boolean Algebra*

1
Computer Science Versus Automation Programming

If you were to ask an everyday automation professional what computer scientists are, you would probably get an answer along the lines of math nerds, computer geniuses, and so on. Most automation professionals are usually engineers or trade persons who stem from a field with little to no computer science exposure. This means very few automation professionals are classically trained in computer principles. For many automation professionals, there is a barrier between complex computing and automation.

This book is going to be different from most automation programming books on the market. This book is going to focus on developing software for **Programmable Logic Controllers** (**PLCs**); however, this book is designed to teach you, the reader, to be more than a PLC programmer or tech. This book is designed to turn you into a genuine software developer. In short, this book will cover everything from program design to security. This book will also utilize Structured Text over the more traditional Ladder Logic. The reason for utilizing Structured Text is twofold. First, Structured Text is the future of PLC programming. As PLC applications become more advanced, the programming apparatus will need to be more robust. Structured Text offers much tighter control over a program than Ladder Logic. Second, implanting a well-designed program in Structured Text will be much easier to implement than it would be in Ladder Logic. With that said, there are no prerequisites for this book. You do not need to have any special math skills, logic skills, or anything of the sort to follow along. Those skills will be introduced in the book, but they will be easy to master and implement.

To begin, the brain of most modern machines is a PLC. PLCs, for many, are just programmable devices, but they are miniature computers that are governed by the same laws of computing that govern any other device, such as a personal computer or smartphone. This poses a problem in the automation world because software is usually considered an easily replaceable component that exists to complement the hardware. In other words, many automation engineers often have an "if it works it'll do" attitude towards software. This is a faulty philosophy, as poorly written software can hinder a machine and put it in the cyber trash heap before its time. Therefore, to become more than a simple PLC programmer, a mastery of computer science is a must.

To begin our journey into computer science, we're going to first explore the following topics:

- What is computer science?
- What is automation programming?
- Why is computer science important in automation programming?
- Why should automation programmers care about automation programming?
- The differences between a PLC and a microcontroller

Finally, to round out the chapter, we will explore the differences between a PLC and a computer.

Technical requirements

This chapter is theoretical and will not require any specific software.

What is computer science?

Computer science is the study of computer systems, with a strong emphasis on software. In a more lay sense, computer science is the study of computational systems such as computers, phones, or anything that runs software, including PLCs. In short, the scope of computer science usually encompasses fields that involve software development or computer architecture. Computer science is a broad field that ties into many other disciplines, such as the following:

- Software engineering
- **Artificial intelligence (AI)**
- Networking
- Cyber security
- Database systems
- Bioinformatics
- Distributed computing
- Computer architecture

The field of robotics and automation can also loosely be considered a field of computer science.

As can be seen, computer science encompasses a lot of different disciplines. For some, this may seem scary, but rest assured that a mastery of each of these subfields is not necessary to be successful at computer science or programming. The focus of this book is going to be mostly on software engineering, which means there is going to be a heavy emphasis on software design and implementation. So, why should one learn computer science?

Why study computer science?

The ultimate goal of computer science is to create faster and more powerful computer systems that can solve increasingly complex problems. To put it briefly, a person would want to study computer science to build more efficient hardware, software, and networks, improve computer system security, and more. In other words, a person will study computer science to build faster, smarter, safer, and more reliable systems. With that, why should automation programmers or engineers care about computer science?

Before the benefits of computer science can be appreciated for automation programming, it is important to understand what automation programming is. As such, the following section is going to explore what automation programming is and where it is used.

What is automation programming?

Automation programming can take on many different interpretations depending on the context and industry. For this book, automation programming will be considered industrial automation programming. Industrial automation programming and control programming can be considered the same thing. When one mentions controls or automation programming, they are usually referring to writing software that lives on some type of controller that is used to automate the use of machinery. In all, automation software is designed to reduce the amount of human intervention in a process.

Automation programming starts with a programmable device. There are many types of automation controllers, with some being the following:

- PLCs
- **Remote terminal unit (RTU)**
- **Proportional – integral – derivative (PID)**
- Miscellaneous control boards

What is considered automation software should include more than software that simply lives on controllers. This means that what is considered industrial control software can also branch out into other families of software, such as the following:

- **Human–machine interfaces (HMIs)**
- **Supervisory control and data acquisition (SCADA)**
- Databases

When someone mentions automation programming, they are usually referring to software that lives on the most common types of industrial controllers, PLCs. PLCs are often the main type of controllers for industrial applications and are often seen as the backbone for many machines. This means that when it comes to automation programming, knowing how to program a PLC effectively is necessary.

For this book, automation programming refers to writing software for PLC devices. With that, before the links between automation programming and computer science can be fully appreciated, it is important to establish a high-level understanding of what a PLC is and what it does.

What is a PLC?

A PLC is a specialized programmable device designed to be very rugged and operate for extended periods. PLCs are responsible for operating industrial or heavy equipment and are commonly used in the following:

- Streetlights

- Amusement parks

- Factories

- Cranes

- Nuclear reactors

- Space launch systems

- Dams

Anywhere a program is required to operate a piece of industrial or heavy equipment, a PLC will usually be present.

For beginners and even some experienced automation professionals, all this computer science stuff may seem dubious, unnecessary, and more trouble than it's worth. However, before any rash judgments are made, let us explore why automation engineers need to understand computer science.

Exploring automation through computer science

In automation, software is often seen as a second-class citizen to the hardware. If you speak to an automation professional, chances are they are going to tout the hardware as the main focal point of the system. Automation engineers love to brag about the latest controllers that are being utilized, how they integrated the finest motors and motor drives into the system, and so on. However, it is rare to hear a typical automation engineer brag about the efficiency of the software or tout the design patterns they used to architect the software.

A lot of this attitude towards software boils down to tangible assets. Often, an engineer can hand a customer the latest power supply or brag about how easy it is to swap out a new motor drive. If the customer ever sells the machine off, they can use all those features as selling points to raise the value of the machine. However, there is a major flaw in this logic. Without quality software, the machine will be an expensive paperweight. With low-quality software, the machine will be a high-quality paperweight that can move and perform certain tasks marginally well at best. Put bluntly, the quality of a machine starts with the software. A machine can have the most advanced hardware in the world, but if it has poorly written software, it will be a poorly performing machine at best. With that, how does computer science help?

How does computer science help automation programmers?

If you think about the way computer science was defined in the past, it was mainly concerned with producing quality software. Computer science has many principles that, when followed, will produce fast, safe, and reliable software. In other words, the computer science principles that are going to be explored in this book are going to allow you to get the most out of that advanced hardware. If you can understand and even master the principles that will be explored, your code will be light years ahead of your competition, and all that fancy hardware will be used to its full potential.

Many experienced automation professionals may be wondering why they should care about computer science and how it relates to automation programming. After all, automation software, for years, has been written with little concern given to software execution performance. There are also a lot of beginners who are more infatuated with robots and building smart factories than with writing quality code. So, to give some context as to why someone should care about computer science, we are going to explore a few reasons why.

Why should automation programmers care?

The world is changing rapidly, and the computer industry is leading the charge. For those who work in the automation industry, where systems can easily be 20 years old, it can often be difficult to see how rapidly the IT world is morphing. In the past, most of the then-emerging technologies were decentralized and did not factor into industrial automation. However, with the widespread adaptation of the internet and interconnected devices, that all changed.

Recently, the world has seen the rise of things such as cloud computing, the IoT, machine learning, and many other things. The new elements have exploded so fast that they are starting to be integrated into the industrial automation realm. To understand why automation programmers should care about computer science, we first need to understand what these modern technologies are. Now, it is important to remember that this list is not exhaustive, but the following technologies will give an insight into why we need to care about computer science principles.

Cloud technologies

The term cloud has been making the rounds over the past few years. In a very lay sense, the cloud is a bunch of interconnected data centers where you can rent resources. These resources vary and have many different applications. For example, some common, high-level services include the following:

- **Virtual machines (VMs)**
- Data storage, such as databases and cloud storage

- Containerization

- Microservice support

- Machine learning

- Data analytics

- Networking

The best way to conceptualize the cloud is an all-in-one resource that has all the computing infrastructure needed to power your application. There is a lot to understand about the cloud and its various levels that go well beyond the scope of this book. However, it is important to know the most popular **cloud service providers (CSPs)**:

- AWS (Amazon)

- Azure (Microsoft)

- GCP (Google)

- OCI (Oracle)

- IBM Cloud (IBM)

It is important to note that this list is not complete, and from this list, AWS and Azure are the most popular.

Most of the services offered through these CSPs charge either via usage, the number of requests to the service, or the amount of time used. What is also interesting is that the cloud is being adopted by the automation industry because it is often cheaper and requires less skill to create resources in the cloud than having to create those resources in-house and maintain the service on a custom server. Even if the automation company you are working for does not utilize the cloud, it is likely that a customer would require the machine to be interfaced with cloud services. With that, the next major technology that we need to explore is the IoT.

The internet of things

The **Internet of Things (IoT)** is another buzzword that has been popping up recently. To summarize, the IoT is a group of devices that are networked together to form an integrated smart system, such as a smart house or smart factory. Typically, devices are connected via a network and can pass data freely to any other device in the network, which allows for easy access to real-time production data. The IoT allows for the following:

- Problems can be identified more rapidly.

- Personnel can have real-time status updates.

- Machine(s) can adjust to changing situations more readily.

- Processes can be better orchestrated.

Overall, the IoT is becoming a very prominent tool in automation and is the backbone of smart factories. Although the IoT is the backbone of many smart factories, another emerging technology is AI and **machine learning (ML)**.

Machine learning

Of all the buzzwords, machine learning is by far the most famous. Machine learning and AI are all over the news, and systems such as ChatGPT are quickly changing the world. Much like the rest of the modern world, AI and ML have infiltrated the realm of automation. Currently, there are libraries that can be utilized to give PLCs the ability to leverage the power of AI and ML. AI and Machine Learning is a complex field that incorporates aspects of mathematics and computer science to understand and properly implement. Additionally, AI and ML are not singular concepts; instead, machine learning and, by extension, AI encompass many different algorithms that do different things. Common algorithms include the following:

- Deep learning algorithms that mimic the human brain

- Regression algorithms that are used to make predictions

- Clustering algorithms that cluster things into groups

These are just a few broad types of algorithms. There are other types of algorithms, and many of those algorithms have different categories. For example, regression algorithms can be **simple regression algorithms**, **multiple regression algorithms**, or **logistic regression algorithms**. These algorithms can open vast new avenues that many would never have dreamed of; however, to effectively use ML, a developer must select a machine learning algorithm, collect and process the data, train the model, and finally deploy the model. This can be an exceedingly challenging job, and depending on the algorithm, formatting the data can be a daunting task that requires a lot of knowledge in computer science. So, what does all this equate to?

What does this mean for automation engineering?

So, with all this, why should an automation engineer care about computer science? After all, automation engineers have spent decades doing just fine with simple Ladder Logic and little thought into the guts of software, so why should automation professionals care now?

Well, the answer is quite simple. Effective programming must be implemented to cut costs, secure the system, and ensure that the system runs efficiently. In terms of the cloud, if a machine is constantly uploading useless data to the cloud, it is going to add an extra cost to the system's operation. For example, if the data is being uploaded to the cloud and the cloud is utilizing services such as computer instances and data resources, each useless byte of data is going to compound the cost of operating the machine. In terms of ML, if those data are being used to train a machine learning model, it can create errors in the model that make it perform poorly. It may also cause the system to take too long to train, which could render it useless. In terms of the IoT, if poor security practices are baked into the system, it can cause the system to become vulnerable to cyberattack. This means that if something is not coded correctly, whole smart factories could be rendered inoperable by simply exploiting something akin to a sensor or voice controller. More than anything else, following computer science principles will simply create better, more robust, and more durable machines. In all, even if you do not use innovative technologies, such as ML, the cloud, or the IoT, it will still pay to learn the core of computer science.

Overall, computer science principles will help secure systems from attack, create smarter systems, and, most importantly, save money. The core of this stems from good coding practices, quality logic, and learning how to implement efficient software. So, now that we know why we need to understand computer science, let us take a step back and look at what a PLC is at the application level. For beginners, especially those who have programmed microcontrollers in the past, the concept of a PLC may not seem necessary. However, in terms of applications, there is a vast difference between microcontrollers and PLCs. With that, let us explore how they differentiate.

PLCs versus microcontrollers

Over the past 15 years, microcontrollers have become quite common for electronics education and hobbyists in general. At first glance, a PLC is a lot like a microcontroller, and if you were to compare a microcontroller to a PLC from 30 years ago, a PLC basically was a rugged microcontroller. Based on the PLC definition established before, it is quite easy to confuse the two types of controllers or, at the very least, confuse the applications. The nature of PLCs has changed, and the two types of controllers are worlds apart. Therefore, it is especially important to understand the differences between the two types of controllers, as it is often more appropriate to use one type over the other. To understand how a PLC differs from a microcontroller, let us explore what a microcontroller is.

What are microcontrollers?

A **microcontroller** can best be thought of as a low-level computer on a chip. Common microcontrollers include the following:

- PICs
- AVRs
- Arduino
- BASIC Stamp
- PICAXE
- Raspberry Pico

It is important to note that many of these devices, such as Arduinos, BASIC Stamps, and Raspberry Picos, are not microcontrollers in the same sense as a PIC. These devices are best thought of as developer boards since they come packaged with a lot of external hardware that AVRs and PICs do not have. It is also important to know that these devices, especially the PICAXE, come with a bootloader loaded on them to understand their special programming interfaces, such as the Arduino programming language or PBASIC. For this discussion, we are going to keep things high-level and group all those devices into the microcontroller category.

In terms of architecture, a microcontroller will often come packaged with its RAM, ROM, CPU, and other peripherals baked onto the chip. Microcontrollers do not run operating systems and can only run one program at a time. In short, most microcontrollers are best described as embedded components. So, what are some of the surface-level differences between a PLC and a microcontroller?

Surface-level differences

On the surface, PLCs and microcontrollers are vastly different. A PLC is a self-contained electronic computing apparatus that will usually have built-in programming connectors and expansion ports. On the other hand, a true microcontroller, such as a PIC, is a chip that requires external components, such as external clocks, resistors, and so on, to operate.

It can be said that PLCs have more in common with Arduinos and other development boards that are microcontroller-based. For example, most development boards can be programmed with a USB or ethernet cable and do not usually require external components other than a power supply to run. However, the similarities usually end there. A major difference between a PLC and a microcontroller is the programming system that they use.

PLC and microcontroller programming languages

In general, microcontrollers are programmed in a derivative of a traditional programming language, typically a C, Assembly, Python, or BASIC dialect. For the most part, each microcontroller will have a corresponding programming language. For example, Arduinos use the Arduino programming language, which is a variant of C++, PICAXE chips use a version of BASIC, PIC microcontrollers use C, Parallax's BASIC Stamp uses BASIC, and so on.

PLCs, on the other hand, are programmed using **Ladder Logic**, **Structured Text** or other dedicated PLC programming languages. Most PLCs are programmed in Ladder Logic, which is a programming language designed to simulate relay logic and digital circuits; however, recently, Structured Text (a text-based programming language that is reminiscent of a cross between BASIC and Ada) has been on the rise. Ladder Logic is, right now, more popular than Structured Text, but when it comes to architecting quality and secure code, Structured Text is much easier to use.

There is also another major difference between PLCs and microcontrollers: standardization. Many microcontroller programming languages do not follow any specific standards. Different programming systems can be used to program microcontrollers if the chips have similar architecture, but the languages themselves will usually vary. Many major PLC brands, such as Allen Bradley, Siemens, Beckhoff, and so on, follow the IEC 61131-3 standard. Essentially, this standard is a set of programming language rules that compliant devices follow. The goal of standardizing PLC programming languages, among many other reasons, is to ensure there is not a drastic learning curve between manufacturers. However, much like microcontrollers, a program written for one PLC device will rarely be able to run on a device made by another company or even another PLC model. With that, how are the two devices used?

Use cases

The use cases for PLCs and microcontrollers are also radically different. For the most part, a microcontroller or microcontroller development board is used for low-voltage applications, with very few designed to give outputs over +5v. Microcontroller chips are normally used in the following types of applications:

- Toys
- Hobby projects
- Consumer electronics (appliances, TVs, etc.)
- Vehicles

Just about anything that is low voltage in nature and is not designed to perform in a complex industrial environment will usually be controlled by a microcontroller. The best way to think of a microcontroller is as an embedded device.

As stated before, PLCs are used for industrial purposes. For example, PLCs are a mainstay in factories because they run for prolonged periods of time without needing maintenance. They are also designed to control high-voltage systems that are usually associated with machinery. A PLC is not an embedded device. On the surface, a PLC can be thought of as an industrial microcontroller, but it is important to realize that they are different from microcontrollers.

Compared to a microcontroller, a PLC is much more rugged and can run for years without needing maintenance and even longer without needing to be replaced. It is not uncommon for a PLC to be in service for decades. PLCs are specifically engineered to be rugged devices that can withstand an extreme environment without interrupting the process(es) they are programmed to control. Though microcontrollers are often used as components to make a PLC, a PLC is not a microcontroller.

This does not mean that microcontrollers are excluded from the world of automation. Microcontrollers are used in many facets of automation, with applications ranging from machine controllers to the chips used in PLCs. Microcontrollers are particularly important electrical components, and their importance cannot be understated; however, to reiterate, they are not PLCs and should not be employed to do the same job as a PLC.

Therefore, a PLC has more in common with a traditional computer than a microcontroller. To appreciate the complexity of a PLC, let us explore the differences between a PLC and a computer.

PLC versus computers

Over the past 20 or so years, the cost of computers has dropped significantly. This drop in price has given rise to cheaper and more powerful computers. This increase in computing power and decrease in price has been reflected in all aspects of life. For example, smartphones are now an integral part of society; virtually everyone has easy access to tablets, such as Kindles, smart homes are on the rise, and more. This shift in computing has also seeped into the automation world. In contrast to the past, hardware-based control panels have been replaced with touchscreen HMIs, advanced networking technologies are employed everywhere, and, of course, PLCs (that more resemble modern computers as opposed to microcontrollers) are a mainstay in most factories. But before the similarities between a PLC and a computer can be explored, it is important to understand what a computer is.

What is a computer?

Describing a computer can be a bit difficult because many devices, such as modern smartphones, have significantly more power than a state-of-the-art desktop computer from 20 years ago. As such, many devices qualify as computers, even microcontrollers, to an extent. This means that many different devices can be considered computers, and whole books can be dedicated to defining what a computer is. However, for our purposes, we are going to keep things broad and think of a computer in terms of a modern, traditional personal computer – that is, a microprocessor-based device that utilizes an operating system and can run multiple applications at the same time.

PLC versus computers

As said before, although PLCs are often conceptualized as industrial microcontrollers, that is only a superficial comparison. Remember, a microcontroller can best be thought of as a low-level computer on a chip because it will often come packaged with its own RAM, ROM, CPU, and other peripherals baked onto the chip. Microcontrollers do not run operating systems and can only run one program at a time.

In contrast, a computer is a microprocessor-based system with external components, such as ROM, RAM, and other external peripherals. A computer system also requires an operating system such as Windows, MacOS, or a Linux distro. When one compares a computer to a microcontroller, major differences start to manifest. For example, a computer's main goal is not to control external hardware in the same sense as a microcontroller does; a computer has way more computational power and can run multiple applications simultaneously.

If one were to examine a PLC, one would find that most of the major brands are microprocessor-based, have external peripherals, such as ROM and RAM, and many of the more advanced PLCs use some form of operating system. For example, a high-end Beckhoff PLC will run Windows, whereas brands such as Wago usually use a Linux distro for embedded devices. The reason PLCs are often touted as industrial microcontrollers stems from lower-end PLCs that behave more like microcontrollers than computers and because the PLC can only run one control program at a time. However, advanced PLCs, such as Beckhoff PLCs, blur the line between PLC and computer, run an embedded version of

Windows, and have Intel microprocessors. On top of that, the more advanced PLCs can usually house and run their own HMI, security software, control program, and so on simultaneously.

Summarily, it is true that a modern PLC behaves similarly to a microcontroller, considering that it can run one control program and, like a microcontroller, controls external circuitry; however, that is where the similarities end. In terms of modern PLC programming and automation programming in general, it is no longer wise to think of PLCs and other control devices as simple microcontrollers. It is better to think of a modern PLC as a computer designed to control machinery.

Summary

This chapter explored the basics of how automation engineering relates to computer science. Thus far, PLCs, computers, and microcontrollers, their use cases, and more have been explored. By this point, a solid foundation on the concepts should have been established.

The days of hodgepodge Ladder Logic programming are ending. The automation industry is notorious for being many years behind the curve in terms of technological advancements; however, over the past 10 or so years, technology has made leaps and bounds that cannot be ignored. With the introduction of technologies such as the IoT, the cloud, and machine learning, automation developers are going to be forced to abandon the old mindset of "if the program works, it'll do." Automation programmers are going to have to adapt to new trends, which means a core understanding of software will be vital in the coming days.

With all that said, a basic understanding of PLC hardware is required to move forward. Just as hardware is nothing without software, software is nothing without hardware. Integrating PLC modules and other hardware is pivotal to the success of a project. As such, the next chapter is going to explore basic PLC hardware components.

Questions

1. What are three use cases for a PLC?
2. Can a PLC be used in a space launch system?
3. Why is computer science important to an automation programmer?
4. What are two use cases for a microcontroller?
5. Name three emerging technologies for industrial automation.
6. What is computer science?
7. Why should automation programmers care about computer science?
8. Name three common microcontrollers.
9. What are some common microcontroller programming languages?

PLC Components – Integrating PLCs with Other Modules

Software and hardware have a symbiotic relationship. Without hardware, software is useless, and vice versa. A key component of computer science is hardware engineering, and as most automation programmers know, hardware is a major factor in automation machinery. In other words, you cannot have a quality machine without quality hardware and software.

In terms of hardware, a PLC is a series of integrated modules that take in data and provide some type of output. To do this, PLCs utilize many different components, such as switches, sensors, motor drives, and safety modules, that all work in unison to control a machine. The key to these systems is the PLC. However, many inexperienced automation engineers do not fully understand what a PLC is, how to integrate external modules, and so on.

Knowing what a component does and how to integrate a module into a PLC system is vital to successful engineering. Even if a person is a dedicated programmer, an in-depth knowledge of hardware is required to both program and troubleshoot a machine. As such, the following is going to be explored in this chapter:

- PLC types
- Common PLC modules
- Sinking vs. sourcing
- Sensors
- Motor control
- Communication protocols
- Wiring diagrams

To round out the chapter, we are going to design a theoretical system that will turn on a motor when a switch is turned on and shut the motor off when a sensor is tripped.

Technical requirements

This chapter will not utilize physical hardware or software. However, to complete the final project, rendering software will be needed. For the most part, no electrical symbols will be used. This means that any rendering software can be utilized, even something as simple as Paint. However, the examples are going to be drawn in block diagram format using `draw.io`, which is free to use online.

PLC types

PLCs are the brains of a machine; however, there is not one single type of PLC. In actuality, there are two major categories of PLCs, which are known as **modular** and **fixed** PLCs. When designing a system, it is important to understand the difference between the two types so the correct one can be chosen for the project. The following is a high-level breakdown of the two types of PLCs:

- **Fixed PLCs**: Fixed PLCs are ones that have their I/O already integrated into them. These PLCs are usually cheaper than modular PLCs but are usually not expandable. This means that whatever I/O comes with the unit is what the engineer is stuck with. Generally, these PLCs are great for standalone or small projects that will never need to be updated. An example use case for a fixed PLC would be something akin to an automatic door opener. For an application like this, there will probably never be a need to expand the system, and only a handful of I/O ports will be needed. In this case, a fixed PLC will be optimal because it will be less expensive.

- **Modular PLCs**: Modular PLCs are more expensive but more flexible. Modular PLCs are very common in large automation projects due to their flexibility and higher memory capabilities. Due to the ever-evolving nature of industrial processes, it is very common to see modular PLCs in factories. Overall, there will be a higher upfront cost with modular PLCs, but they will offer more expandability in the future.

Modular PLCs can be considered to be more complex than fixed PLCs. More thought must go into the system design since modular PLCs are not integrated units. Therefore, to fully grasp modular PLCs, the modules that compose them need to be explored.

Common PLC modules

As could be deduced from the last section, a modular PLC is not a singular unit. PLCs are composed of many different modules that, when integrated together, form the PLC. Many different modules do many different things; however, all PLCs have a select few modules that perform the same functionality across all brands. This section will be dedicated to exploring some of those modules. With that, let us explore the power supply module!

Power supply

As everyone knows, electronic devices require electrical power of some type. PLCs require a stable power supply to function properly. Most PLCs utilize a 24VDC power supply. Depending on the PLC brand, the power supply will either be a module that attaches to the PLC in some way or it will be directly wired to the PLC. Of all the modules, the power supply is probably the easiest to comprehend. Let us now explore what a PLC chassis is.

Chassis

Depending on the PLC system being implemented, modules will need what is called a **chassis** or **rack** to interconnect with each other. A chassis is like a carriage that houses multiple modules and allows the modules to communicate with each other. A chassis is not always necessary, as some devices connect by simply attaching to each other. For example, to integrate Beckhoff modules, all one must do is slide them together and set an address on the device. Whether a PLC requires a chassis or not will depend on the manufacturer and model. With that, we can investigate the CPU.

CPU modules

Regardless of the model, all PLCs have one key component. This component is known as the **CPU module**. Many inexperienced engineers consider this module to be the PLC itself; however, the CPU is actually a module as well. If the PLC is the brains of a piece of equipment, the CPU is the brains of the PLC. This module is the computer that houses the PLC program, memory, operating system (if applicable), and all the other necessary components that are needed for the program to tick.

This is the main module an engineer or tech will interact with. These modules usually have some type of communications interface, such as a USB port, Ethernet, or other custom port that is designed to interface with a standard personal computer that contains the programming software. By interfacing with the CPU module, an engineer or tech will be able to do the following:

- Upload/download a program
- Alter the program
- Change values in the program, such as timer and counter setpoints
- Read the real-time data collected by the unit

The CPU module will typically be integrated with other modules, such as the I/O (input/output) modules that will send signals to it. The CPU will then process those signals and carry out the programmed instructions. If the CPU module fails, the system will fail with it. Now, if the CPU is the brains of the PLC, the I/O modules can be thought of as the nervous system.

I/O modules

I/O modules are pivotal to a PLC as these modules are responsible for receiving inputs and driving outputs. In terms of I/O modules, there are two main categories that engineers need to know about. The first category is digital, and the other is analog. Both categories have their own uses, and both are of equal importance. Most PLC systems will incorporate a combination of the two types of I/O, which means that an engineer won't be able to function without a basic knowledge of how they both work. Therefore, to begin the discussion, let's explore digital modules.

Digital I/O

Digital I/O modules, or as they are sometimes called **Discrete I/O modules**, are PLC modules that operate in either an on or off-state. Digital I/O modules fall into two categories: inputs and outputs. An input device is usually tied to something like a sensor or switch and once it detects voltage over a certain threshold, it will send a signal to the CPU module that the data point is on. Once the voltage is below a certain threshold, it will lose the signal to the CPU module, therefore letting the PLC know that the data point is off. On the other hand, output modules are usually tied to devices such as relays, LEDs, alarms, and so on. Output modules are used to control the state of these devices. In other words, the CPU module will send a signal to the output module telling it to turn the device on or off.

Overall, it is important to know that discrete devices are like switches; they are either fully on or fully off, and there is no in-between state. Though threshold voltages and outputs can sometimes vary from module to module, 24V will usually trigger a digital module to the on state, and an output module will usually be 24V when activated. For many cases, having a fixed on or off state is not desirable, so an engineer can employ analog modules for cases like these.

Analog modules

Analog modules are like digital I/O modules in that they send and receive signals from the CPU. Unlike their digital counterparts, analog I/O can send and receive a wide range of current or voltage levels. Since analog I/O can send and receive signals of varying sizes, they are used with components that either send or receive various levels of voltage or current.

In terms of analog inputs, these modules are often used with sensors that will return a value based on a corresponding physical state. That state will produce a signal proportional to that state. Common examples of analog sensors that are often integrated with analog inputs are pressure sensors, thermocouples, strain gauges, current/voltage sensors, and the like.

Analog output modules work similarly to analog inputs but in reverse. Instead of receiving current or voltage, they produce current or voltage. These modules are often used in conjunction with devices that use a variable electrical signal to change their physical state. Common devices that are used with analog outputs are motor drives, valves, heating coils, power supplies, and the like.

Safety modules

Automation can be a dangerous field. If a component fails, it can cause unsafe conditions for both the machine and, more importantly, the people around the machine. To ensure a safe and graceful failure, engineers often incorporate safety modules into the PLC design if the base modules are not compliant with the IEC-61508 safety standard.

Safety modules are modules with redundancies and self-checking functionality that ensure nominal behavior. Much like traditional modules, there are a multitude of safety modules. Typically, these modules are used to integrate safety devices into the system. For example, sensors (more information on sensors is provided in the sensor section), such as safety sensors and emergency stops, are often wired into these modules. As such, if things like light curtains, trip sensors, or any other safety sensor or stopping mechanism are integrated into the system, it is best to wire these devices into a safety module.

The world of modules is very rich, and there are many different types that have not been explored here. The next step in understanding modules and, by extension, PLCs is to understand the difference between sinking and sourcing signals.

Sinking versus sourcing

In automation engineering, the concepts of sinking and sourcing are very important. Sinking and sourcing is, in lay terms, the direction of the current flow. For example, consider *Figure 2.1*:

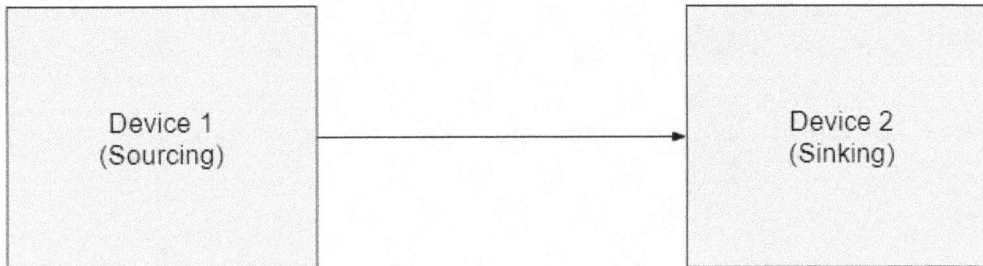

Figure 2.1 – Sinking and sourcing

In *Figure 2.1*, the arrow that represents the current is going from Device 1 to Device 2. This means that Device 1 is the sourcing device, and Device 2 is the sinking device. If the arrow is reversed, then Device 2 would be sourcing, and Device 1 would be sinking. In other words, this concept boils down to the direction of the current flow.

NPN versus PNP

Another way to think of sinking and sourcing is to consider an NPN or PNP device. For many who have studied electronics in the past, PNP and NPN may seem very familiar because they are the two basic types of transistors. The easiest way to conceptualize NPN and PNP devices is as follows:

- **PNP**: A PNP device is a sourcing device. This means that the device is placed between the positive voltage rail and the load.

- **NPN**: An NPN device is a sinking device. This means the device is placed between the load and the negative voltage rail.

To visualize this concept, consider *Figure 2.2*:

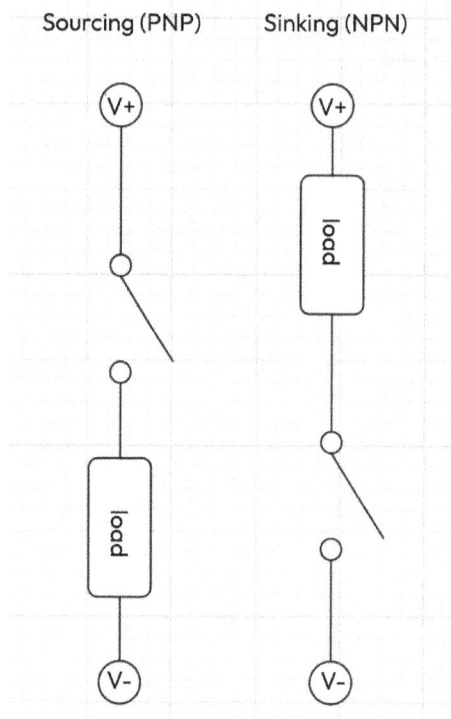

Figure 2.2 – Sinking vs. sourcing

Digital I/O devices are either designed to be sinking or sourcing. It is very important to understand what type of device is being employed. As an automation engineer, you'll mostly encounter sinking inputs and sourcing outputs. This means that things like sensors are usually PNP in nature, while modules are NPN. There is a "gotcha," though; older PLCs are inverted and may require NPN-style sensors. Now that some background information for sinking and sourcing has been established, it is time to explore sensors.

Sensors

If one were to compare a machine to the human body, a sensor would be something akin to nerves. A **sensor** is a device that takes in some type of input from the outside world and converts it to an electrical signal that a PLC module can read and transmit to the CPU module.

Generally, sensors will produce an analog value or an on/off value. An analog sensor is a device that will produce an electrical signal in proportion to its reading. For example, a light sensor, such as a photoresistor, will produce an electrical signal that is proportional to the amount of light that is picked up. In a more lay sense, an analog device is one in which the output signal will vary. Common analog sensors are as follows:

- Light sensors
- Strain gauges
- Sound gauge
- Pressure sensors
- Position sensor
- Temperature sensors

On the other end of the spectrum are sensors that produce an on/off value. These types of sensors are like switches. An example of a digital signal might be an IR sensor that is used to detect entry into an area. If the light beam is not broken, it will produce an on signal that will be picked up by the digital I/O module and transmitted to the CPU. If someone were to break the light beam by stepping in front of it, the digital I/O module would detect an off case, which would be transmitted to the PLC. Common digital sensors are sensors that will produce an on or off state, such as a switch, IR sensors, proximity switches, and so on.

So far, sensory input, modules, and more have been explored. However, what has not been touched on is the main catalyst to turn all those signals into something meaningful: motion.

Motors and motor controls

Circling back on the human body analogy, motors are like the muscles of a machine. Without motors, most machines are glorified paperweights. Motors are responsible for propulsion, moving conveyors, positioning machines, drilling holes, and much, much more. So, fundamentally, what is a motor?

What is a motor?

At a high level, a **motor** is simply an electrical device that converts electricity to circular or linear motion. There are many types of motors, such as the following:

- DC motors
- AC motors
- Servo motors
- Stepper motors
- Linear motors
- Brush/Brushless motors

Automation utilizes a vast majority of motors, and each type of motor will have its specific use cases. In terms of automation, arguably, the two most commonly used motors are steppers and servo motors. Before stepper and servo motors can be explored, it is first important to understand the difference between open and closed-loop control.

Open and closed-loop control systems

The terms open and closed-loop control systems may sound a bit intimidating at first, but the concepts are very simple to understand. In a very basic sense, open and closed-loop controls simply mean whether or not position data are fed back into the control systems. The two types of systems can be summarized by the following:

- **Open-loop control**: No positional data is fed back into the control system
- **Closed-loop control**: Positional data is fed back into the control system

Both open and closed-loop systems are common in automation. However, depending on the type of motor involved, extra hardware might be required, such as an encoder or resolver. Now that a little background information on open and closed-loop systems has been established, we can move on to exploring stepper and servo motors.

Stepper motors

Stepper motors are high-precision motors. Stepper motors are often found in common devices, such as printers, 3D printers, hard drives, and so on. If an everyday device has a motor, it is probably a stepper motor. Unlike traditional motors, stepper motors can be positioned into a number of equal locations. As the name suggests, a stepper motor will move into what is known as a step. A step is the number of degrees the motor shaft will move in response to a signal. The amount the motor moves in response to a control signal is called resolution or the step angle. The number of steps for the shaft to complete one full revolution can be found with the following:

$$Steps = \frac{360}{Step\ Angle}$$

For stepper motors, control signals are pulses. Generally, stepper motors will move 1.8 degrees for each pulse and will usually require 200 steps for a full rotation.

The number of pulses required to move the motor to a specific position is given by the following equation:

$$Pules = \frac{Degree\ Rotation}{Step\ Angle}$$

So, if an engineer needs to position the motor 45 degrees, and the motor has a step angle of 1.8 degrees, the engineer will need to use the following equation:

$$Pulses = \frac{45}{1.8} = 25$$

This case will require a total of 25 pulses to the motor for it to move 45 degrees. Since no positional data are fed back into the system, stepper motors are generally considered open-loop control systems, but it should be noted that with the proper hardware, they can be converted to closed-loop systems.

Stepper motors are used in applications that require precise positioning but slow speeds. Typically, stepper motors are used in applications that require low speeds, as this is how stepper motors usually perform best. Common applications for stepper motors are devices such as ground-based telescopes, antennas, robots, turntables, and so on.

Servo motors

Another very common motor used in automation is the **servo motor**. Servo motors are the inverse of stepper motors, as they perform better at higher speeds, which allows them to be used with gearboxes to deliver more torque. Unlike stepper motors, servo motors operate under a closed-loop control system. This means that servo motors require a type of sensor called an encoder to provide feedback data to the control system. Though technically optional, servo motors typically require a motor drive and an encoder to function as intended. With that, let's explore encoders.

Encoders

By default, not all motors have mechanisms to determine their speed and position. Unlike stepper motors, servo motors need a mechanism to collect data on the motor's position, angle, and so on. **Encoders** are basically sensors that collect data on a motor's movement. Encoders are typically used with servo motors; however, they can also be used with stepper motors as well. Generally, encoders can be changed out; however, most servo motors will usually come pre-assembled with a built-in encoder that is specifically designed to work with both a motor drive and servo.

Motor drives

A pivotal component of any automation system is a motor drive or **variable frequency drive (VFD)**. In a lay sense, a **motor drive** is an electrical device that sends control signals to a motor. These devices are used to control the motor speed, torque, and position of the motor. There are many different types of motor drives that are used to control the various types of motors. The four main types of motor drives are the following:

- AC drives
- DC drives
- Servo drives
- Stepper drives

The way each one of these drives operates is different and goes beyond the scope of this book. However, each of these drives will take control inputs from a controller device such as a PLC.

Motor drives need a way to communicate with a PLC. A PLC and a drive can communicate in various ways, depending on the type and application, with arguably the most common being some kind of communication protocol.

Communication protocols

Communication protocols are an advanced concept in automation. Consequently, only a high-level explanation of communication protocols and how they work will be given here. However, communication protocols, along with more advanced software engineering concepts, can be explored in my previous book, *Mastering PLC Programming: The software engineering survival guide to automation programming*.

What is a communication protocol, and what is it used for?

Communication protocols are a way for devices to pass data to and from each other. In the most basic sense, a communication protocol is a way of packaging data in a common format that devices on a network can understand. There are many different protocols that are used in automation engineering. Some protocols are common protocols that are used throughout the IT industry, while others are proprietary and designed to work only with specific control systems. Common communication protocols are as follows:

- UDP
- TCP/IP
- Modbus
- EtherCat
- Profinet

- Profibus
- MQTT

From this list, UDP and TCP/IP are used widely in all industries, including internet communication. The other protocols are mostly used for automation systems, such as PLC-based systems, among other things. Now that networking, drives, and modules have been explored, we can move on to wiring diagrams!

Wiring diagrams

The main way to communicate a design to others is with a schematic or wiring diagram. Diagrams can be as detailed or as simple as needed; all that matters is that enough information is given to relay the necessary information regarding the internal workings of the systems. This book is going to represent systems as block diagrams, in other words, as high-level diagrams. For example, wires and dataflows can be represented with simple lines and components, and software services can be represented as shapes, such as squares. To fully demonstrate wiring diagrams, let's work on the final project.

Final project

Many programs can be used to draw wiring diagrams. For this book, we are going to use **draw.io** to draw diagrams. You do not have to use `draw.io`, so feel free to use a different drawing system or even draw out the diagram on paper if you prefer. It is important to note that drawing diagrams are very subjective. There are many ways to draw a diagram, and, as such, there is no right or wrong way of doing things. All that matters is that the connections are accurately represented on the diagram. Therefore, it is okay if your diagrams do not match the diagram in the book.

This project is going to use theoretical parts. In automation, the way in which devices operate will vary widely. This means that we must make some assumptions about how these parts operate. It is important to note that this project is not meant to produce a working system. Instead, this project is merely an exercise in design logic to get a feel for how components will interact.

Specs

Before the diagram can be drawn, we need to lay out some requirements about how the system should operate:

- A conveyor motor should activate when an IR sensor detects a box
- When no box is detected by the sensors, the motor should shut down
- The motor speed needs to be adjustable with the turn of a knob

- There should be a light curtain that should shut the motor down when it is tripped
- An **emergency stop** (**E-Stop**) needs to be integrated into the system to shut the system down in case of emergencies

Needed components

From the description, we're going to need a digital I/O module to handle the IR sensor. Since the requirements specifications only require one sensor and digital inputs usually can handle multiple inputs, only one digital input module will be needed.

In terms of the speed control knob, a simple potentiometer (variable resistor) can be used. A potentiometer will be tied into an analog input. As the resistor is turned, it will vary the voltage to the analog input module, which can then be processed by the CPU.

The requirements also mentioned two safety features: the light curtain and the E-Stop. Since these are safety devices, the system will either need safety-compliant devices or separate safety PLC modules. In this case, we're going to assume that the PLC is not a safety-rated PLC system, and we will add safety modules into the system.

The final aspect that needs to be explored is the drive system. For this project, a relatively simple system is all that is needed. For this, we will use a simple DC motor. Since the PLC is controlling the system, we cannot connect the motor directly to the PLC, especially since we need to vary the speed. This means that we will also need a DC motor drive. Additionally, most drives do not connect directly to the PLC and, instead, use some intermittent module, such as a contactor, that can be used to relay signals. As such, we are going to create a generic PLC interface that will act as a relay system for communication with the motor drive. Now that we have an idea of what we need to design the system, let us make a **bill of material** (**BOM**) to get a parts list.

BOM

A BOM is simply a parts list for all the materials needed for the project. Engineers will usually produce the BOM after the design is finalized; however, for learning purposes, we're going to create the BOM first so we do not forget any parts:

- Power supply
- CPU module
- Digital input module
- Analog input module
- Safety CPU
- Safety digital input

The sensors comprise the following:

- E-Stop push button
- Light curtain
- Potentiometer (10K)
- IR sensor

The motor system is made up of the following:

- DC motor
- DC motor drive
- Generic motor-PLC interface
- Network cabling

Before looking at the diagram, try to sketch out a mock diagram of what you think the diagram should look like. Assume the following:

- E-Stop on **Safety Digital In 1**
- Light Curtain on **Safety Digital In 2**
- All other devices integrate into **Input 1** of their respective module
- Draw the networking cables as a thick line

Once complete, you should have something akin to the following:

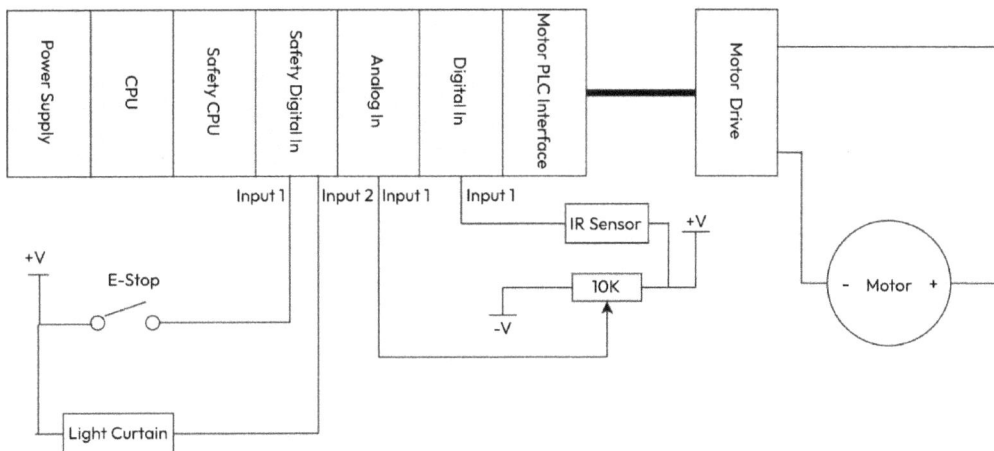

Figure 2.3 – Wiring diagram for system

In this diagram, the wires are represented by thin lines, and data lines are represented by thick lines. In terms of the motor, we're assuming that the motor driver receives signals from the CPU via the Motor PLC interface and will control the motor speed by simply controlling the current to the motor. The way this would work in the real world will vary as different motors and motor drives will operate differently. Some only require the motor to be hooked up to the drive, while others have to be connected to a power source and are controlled via a control signal. It all depends on the motor, drive type, and the manufacturer.

It is also important to note that we are using a safety CPU in conjunction with a regular CPU. This is a common practice when using a non-safety-compliant PLC system. Safety CPU modules usually house their own programs; however, they are usually integrated into the overall system and communicate with the main CPU. The programs that these CPUs house are usually things related to safety inputs, starting dangerous subsystems, and so on.

Summary

This chapter explored the integration of modules and components into a PLC system. The different types of PLC modules, sensors, motors, communication protocols, and more have been explored. We have also designed a simple PLC-based conveyor system to reinforce the concepts that were covered. At this point, you should have a basic idea of how hardware works and how it is used in an automation system.

As stated before, hardware is an important aspect of automation. However, as was also stated, it is not the only aspect that engineers should care about. Now that we have some hardware theory under our belts, we can move on to explore programming concepts by looking at the fundamental principles of programming!

Questions

1. Name three types of PLC modules.
2. What is the difference between a regular PLC module and a safety module?
3. What is a stepper motor?
4. What is a servo motor?
5. How many pulses are required to move a stepper 180 degrees if the resolution is 1.8?
6. What is a motor encoder?
7. What is a motor drive?
8. What is the difference between an analog module and a digital module?
9. What is a discrete module?
10. What is a BOM?

11. What type of module should an E-Stop be wired into?

12. What is an analog input?

13. What is a discrete input?

Further reading

*   **Types of motor controllers and drives**: https://www.thomasnet.com/articles/instruments-controls/types-of-motor-controllers-and-drives/

*   **Industrial motor drives**: https://www.danfoss.com/en-us/markets/industry/dsp/industrial-motor-drives/#tab-overview

*   **Motor encoder overview**: https://www.dynapar.com/technology/encoder_basics/motor_encoders/

*   **What is the difference between NPN and PNP?**: https://www.maplesystems.com/supportcenter/faq?qid=328

*   **Sinking vs. sourcing in PLCs**: https://www.dosupply.com/tech/2022/05/16/sinking-vs-sourcing-in-plcs/

3
The Basics of Programming

For many college students, and even some entry-level automation engineers, programming can be the stuff of nightmares. For the uninitiated, programming is cloaked in a shroud of mystery that only the brightest can peer behind. The media has reinforced this cliché with movies that feature super-genius programmers who are able to do complex calculus in their heads and write a million lines of code a second. However, this is a very fantastical and untrue cliché of what programming in the real world is. This false reality is also very damaging to up-and-coming programmers as it has painted an unrealistic view of the IT world that has discouraged many young developers. Programming doesn't require a person to have any more intellect than any other field. What can be said, though, is that programming requires a bit more patience.

The key to a solid foundation in programming is understanding the basics of how programs work. No matter the language, platform, or application, there are common themes, such as the flow of a program and the way a programming language works. Once these concepts are understood, programming will make a lot more sense.

To dispel some common misconceptions, to be a successful programmer, one does not need to be a genius, gifted at math, able to think in multidimensional space, or anything of the sort. The qualities that one does need are patience, curiosity, and a will to learn. As cliché as it sounds, the best programmers are not always the most intellectually gifted. In fact, some of the best developers are those who weren't very good at academics in school but had a true passion for learning.

The core of PLC programming is the IEC 61131-3 standard. To many, that is a meaningless tag, but to an experienced automation programmer, it says a lot about the PLC that follows it. To fully appreciate the standard, we have to lift the programming veil; to do that, we're going to explore the basics of programming and their languages. As such, this chapter is going to cover the following concepts:

- Understanding what a program is
- Understanding programming languages
- Exploring keywords
- Exploring program execution

- Exploring the flow of a program
- Exploring IEC-61131-3

To round out the chapter, we're going to explore some exercises to help comprehend algorithms.

Technical requirements

This chapter is going to explore programming from a theoretical point of view. Therefore, no special software will be needed. However, a text editor such as Notepad or even Microsoft Word would be beneficial to help write draft algorithms.

Understanding what a program is

The first step in understanding how to write a **program** is to understand what a program is. The technical term for a program is **algorithm**. For the inexperienced, the term algorithm is ambiguous and, for some, scary. However, an algorithm is simply a set of steps. This means the most literal way to think of a program is as a series of steps that a computer or other programmable device will carry out to accomplish a task.

To demonstrate what an algorithm is, consider the steps it takes to withdraw money from an ATM:

1. Enter the debit card into the machine.
2. Enter the PIN.
3. Enter the amount to withdraw.
4. Remove the debit card from the machine.
5. Take cash and go.

The steps it takes to withdraw money from an ATM is a prime example of an algorithm. Now that we know what an algorithm/program is, what purpose does it serve?

What is the purpose of a program?

The purpose of a program starts with a problem to solve. For our purposes, a problem doesn't necessarily mean that there is something wrong. Typically, a problem in a programming sense usually means that there is a task that needs to be accomplished. An example of a problem can be anything such as the following:

- Creating a text editor
- Controlling a robot

- Controlling a drill bit
- Performing calculations

In other words, a program is anything that runs on an electronic device that helps people complete a task and increase productivity. With that, why should people opt to use a program instead of using electronics or mechanical components?

Why use software over hardware?

Software is a non-physical component. Software does not take up physical space in a machine and when designed properly can be easily scaled to accommodate new functionality. Outside of development costs, software carries no extra expense. This is in stark contrast to adding extra and expensive hardware to accomplish the same task.

As stated many times throughout this book, many automation engineers approach problems with a hardware-first mentality. This is a very poor mentality that can drastically increase the cost, size, and points of failure for the project. Adding extra hardware can often solve a problem; however, if a problem can be solved programmatically, it is usually best to opt for that path. So, how should an engineer view hardware and software in a system?

How to view software and hardware in a system

Up until now, this book has posed what many would see as a counter-intuitive view of automation engineering. Usually, automation engineering students and engineers are taught to approach software with a hardware-oriented view; hence, the creation of Ladder Logic, which simulates relay-logic diagrams. However, when viewing the hardware-software relationship, one needs to view software as the workhorse of the system. Software should be in charge of the orchestration of all the functionality of the system, while the hardware only exists to house the software and control the machine. Hardware in a system should serve one purpose and one purpose alone, to support the software. As stated before, this can seem quite counter-intuitive because many academic programs focus on hardware significantly more than software.

Software is not a cure-all solution

Software is not a cure-all. Many automation programmers typically take on another very poor philosophy in that they usually try to compensate for faulty hardware with software. This is not the goal of a program and can lead to many issues down the line. A code base should never be altered to compensate for a broken physical part. The only time a code base should be modified is when a new feature is being added, an old feature is being removed, changing equipment behavior, or a physical part of a process is altered.

There is one exception to this rule and that is when calibrations and similar operations are required. Sometimes, a machine will not have an interface to input data. This is especially true for older machines. In these cases, the only way to input the data is by modifying the values in the program. In cases such as calibration, an engineer is not modifying core logic; they are only modifying certain values in the code base.

Thus far, we have explored what a program is as well as where and how it should be used. We've also explored the general steps of how to outline an algorithm. The one thing we have yet to explore is how to create a program.

Understanding programming languages

The key component of creating a program resides in what is called a **programming language**. A programming language is a special language that both a computer and a machine can understand. There are many different programming languages that are available, especially for PLCs. In terms of general-purpose text-based programming languages, there are a few popular ones, which are as follows:

- Java
- C/C++
- C#
- Python
- JavaScript/Node.js

There are many more general-purpose programming languages available, these are just a few. In terms of PLC programming, there are some specific languages that can be used depending on the PLC being used, which are as follows:

- Ladder Logic
- Sequential Function Chart
- Function Block Diagram
- Structured Text

Now, having so many different programming languages may seem redundant, but each language has its own niche, strengths, and weaknesses, and many of the languages follow different paradigms (ways of structuring code). With that, let's explore the basics of a programming language.

Syntax

Each programming language has a unique set of rules. The rules that govern the language are called **syntax**. The syntax for a programming language is very similar to the syntax of a written language. That is, for the computer to make sense of the program, it must follow a certain grammar construct and sequence. For the most part, each language will have a unique syntax; however, many general-purpose languages follow the general rules of the C programming language. Now, a computer or programmable device doesn't understand written language the same way as humans do. Computers speak the language of 1s and 0s or ons and offs. This means that there has to be an intermittent step to convert the human-readable program that a person writes into a program that an electronic device can understand.

Translators

Obviously, a program has to be translated from human-readable code to a language a machine can understand. This conversion is accomplished in translation. There are different types of translations; however, for everyday use, the two most common are compilers and interpreters. The differences between the two can be summarized with the following:

- **Compiler**: A compiler is a program that reads all the source code at one time and converts the code into commands that a machine can understand. During the compilation process, the compiler will check the human-readable code for errors that would otherwise cause errors in the compilation process and will abort the process if any errors are found. The final output from the compilation process is all the human-readable code converted into machine commands.

- **Interpretation**: An interpreter is another type of program that converts human-readable code to machine code. However, an interpreter works differently than a compiler. Where a compiler converts the source code all at once, an interpreter will convert the code to machine code line-by-line as the program runs. Interpreters are typically used with scripting languages, but that's not necessarily a hard rule. Interpreted programs are generally slower than compiled ones due to the line-by-line translation; however, on modern machines, this is usually not noticeable. Interpreted software can pose more risk during development as fatal errors are not caught until a block of code is run. This means unusual conditions can pose problematic risks with interpreted languages.

Whether or not a program is compiled or interpreted may seem academic; however, it is very important to understand how a program is translated to machine-understandable code. In terms of PLCs, most programming systems utilize compilers; as such, when writing the software, you must ensure that you do not violate the rules of the programming language. Now that we have explored the basics of translators, we need to shift our attention and explore machine commands.

Machine instruction

The commands that a program translates into are called **machine instructions**. Machine instructions are special commands unique to a processor or family of processors. These instructions command the processor to carry out operations such as moving data around and logical operations. Typically, a developer won't have to directly work with the machine instructions, but a well-rounded one should be aware of them.

So far, our attention has been focused on the mechanics of a programming language. However, there is more to a programming language than simply the way it works. The next vital concept that we need to explore is programming paradigms.

Language paradigms

Most programming languages require programmers to organize and structure programs in a specific manner. This way of organizing code is called a programming paradigm. The most common paradigms are as follows:

- Object-oriented

- Functional

- Procedural

- Declarative

- Imperative

This is not a complete list of programming paradigms, but these are very common. Most languages are not pure, meaning they do not completely follow one paradigm or another. Most are a mix of paradigms and tend to cherry-pick the features that best suit the language.

Of all the paradigms, **Object-Oriented Programming** (OOP) is the most common. Almost all modern programming languages support OOP to one extent or the other. In fact, OOP is so ingrained in modern programming that it is a prerequisite for any programmer hoping for employment. In short, object-oriented programs utilize a data structure called a class, or as they are known in IEC-61131-3, **function blocks**. A class or function block is essentially a digital blueprint of an object. This is an extremely important concept; however, this concept goes beyond the scope of this book.

For PLCs, the programming paradigm will either be procedural or object-oriented. The IEC-61131-3 standard supports OOP, but whether or not a PLC manufacturer supports it or not will vary. Common PLC programming systems that support OOP are those that are built off the back of CODESYS.

Understanding which paradigm a programming language uses will seem very academic to many inexperienced readers. However, this is one of the most important aspects that a developer should know as it will dictate the way they structure their program. Though a program's structure is very important, it is also important to understand the concept of keywords to implement that structure.

Keywords

Every programming language has **keywords**. Keywords are reserved words for a programming language that performs certain tasks. For example, keywords are used to declare control statements, function blocks, functions, datatypes for tags, variable blocks, and so much more. In other words, keywords are commands. When a keyword is used, it will signal to the PLC to do something such as compare two numbers or create a variable.

Keywords cannot be used to name variables, functions, function blocks, or anything else; as such, they are referred to as reserved words. Keywords in a decent programming editor will change color to let you know that you are using a reserved word. Keywords will vary greatly from language to language as well as their functionality.

Dos and don'ts of learning keywords and syntax

Keywords and syntax can pose a major pitfall to developing programmers. Many inexperienced developers feel they have to memorize a language's syntax and keywords to be a quality programmer. This is a grave misconception and a stereotype that is often propagated by inexperienced developers and low-quality hiring managers. The key to mastering a programming language and mastering programming, in general, is learning the key concepts that govern the language. In other words, think of learning a programming language the same way that a person would learn a foreign language. If you were to compare a high school student who is learning Latin to a linguist who is learning the language, who do you think would have a better grasp of the language? Hopefully, you would say the linguist who understands the concepts that govern language syntax over the person who is simply memorizing phrases. Learning a programming language is no different. Once you learn the core concepts behind the language, you'll be able to master not only the language you're working on but any language you encounter in the future that follows the same paradigm. Consequently, when you're reading this book or reading programming material in general, don't focus on the syntax or memorizing the pattern, but instead focus on the core ideas behind the concept.

For keywords to make sense programmatically, the flow of a program needs to be understood. With that, we're going to look at the flow of a program.

Program flow

The flow of a program can often be confusing to an inexperienced programmer regardless of whether they are a traditional programmer or an automation programmer. However, the overall flow of a program is very simple. In the most basic sense, a program will flow from top to bottom. A program will start at the very first command and will end at the last command. However, a program can have multiple paths to that last command.

Though a program will flow from top to bottom, the path it takes to the bottom may vary. A program can branch out into different paths, code blocks can be looped over, and functions that live in other files can be called. It is also important to understand that the last command that is executed is not necessarily the last command in the file or group of files. We're going to explore some of these concepts in more detail later on in the book; however, for now, just assume that a program will start at the first command and end at the last command.

Program iteration

For most programming languages, when a program executes its final command, it will automatically terminate; that is, it will stop running and will need to be restarted by a user. Having to restart a program, especially in a high-paced automation setting, is not optimal. Therefore, for applications such as PLC programs, you want the program to run in a loop. This means when the last command is executed, you want it to loop to the top of the program and start over from the first command. Consider *Figure 3.1*:

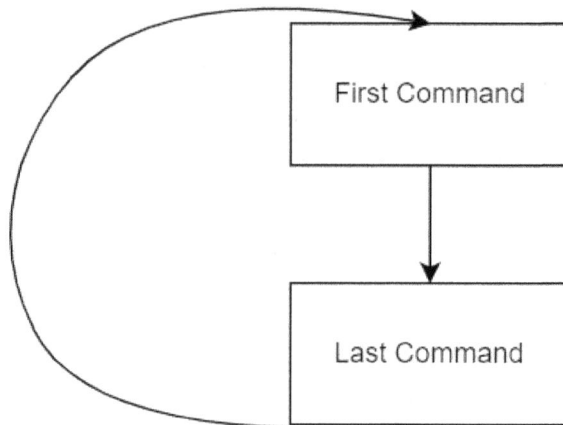

Figure 3.1 – Program flow

Figure 3.1 shows the general flow of a PLC program. Most code blocks can be looped over, but to do that will require a special set of keywords. Most PLC programs, however, will generally run in an automatic loop without the need for any special code.

This is very important to remember as you're developing PLC programs. It is common for many PLC programmers, especially those who have training in traditional programming, to try to create a loop for the PLC program. This additional loop is not only unnecessary but can cause performance issues in the program. Always remember that your PLC program will always restart at the very first command when it reaches the last.

The last relevant concept to explore is the IEC 61131-3 standard. So far, we've explored the guts of the programs and programming languages, but we haven't really explored how that all integrates into the IEC 61131-3 standard. The following section is going to explore the standard and how everything fits together for PLC programming.

Exploring the IEC 61131-3 standard

Most of the major PLC brands follow what is known as the IEC 61131-3 standard. The standard is designed to add programming uniformity across all the brands that utilize the standard. The standard provides uniformity for the following:

- Syntax
- Base keywords
- Language semantics

In other words, when utilizing a PLC that is compliant with the standard, one can expect a similar programming experience when programming any device that follows the standard. This means that a programmer will not have to worry about learning a different PLC language, as the core commands and grammar will be the same. Currently, the IEC 61131-3 standard provides programming protocols for the following PLC languages:

- Ladder Logic
- Structured Text
- Instruction List
- Sequential Function Chart
- Function Block Diagram

Depending on who you speak with, learn from, or work with, the names in the preceding list may differ. For example, the ladder language can be referred to as ladder diagram programming, Ladder Logic programming, or any other variation. Sequential Function Chart is often referred to as sequential flowcharts, or simply flowcharting. Structured Text is often referred to as simply text, and so on. A lot of the name variation is due to the slang and culture of the programmers. Regardless of what the language is referred to, an IEC 61131-3-compliant device will follow the same general rules. Though there will be similarities in programming-compliant devices, there are some pitfalls to be aware of.

IEC 61131-3 pitfalls

The first significant pitfall that one needs to be aware of is that though the standard does create uniformity across the programming languages for compliant devices, it does not mean that programs for one PLC model can be downloaded to another. Generally, software written for one PLC will not port over to another PLC brand or, in many cases, another model of the same PLC brand. Much of this has to do with the way the code is compiled, the architecture of the devices, and so on. This means that though the syntax and grammar may be similar across compliant devices, a developer cannot simply take a program file for one PLC and download it to another.

Another pitfall to look out for is custom functions in the programming environment. It is not unusual for manufacturers to include their own prebuilt code in their programming environment. It is also not uncommon for manufacturers to *lock out* the ability to use certain functions in certain models. This means that if a developer has a program for an IEC 61131-3 PLC and they try to port it over to another device, even if the other device is compliant, it may not work. That means that function may have to be manually programmed in.

The last pitfall to look out for is the features that are included in the PLC programming language. It is important to remember that the IEC 61131-3 standard is just that, a standard. This means that PLC manufacturers can cherry-pick what IEC 61131-3 features they want to include. A common example of this is support for OOP. OOP is a novel concept in the PLC world. As such, many manufacturers have not integrated support for the paradigm yet. This means that if you were to try to port a program that utilizes OOP architecture and keywords to a device that does not support the paradigms, the developer will likely be met with a series of compilation errors. The same goes for any feature in the IEC standard; just because the standards support it, doesn't mean the device will.

A key issue faced by many students who are required to study software development, and even entry-level developers, is learning to think like a machine. Unless a person spent their formative years writing code, it is unlikely that the logical thinking required to write code will come naturally to them. However, there is a simple exercise to help with this that we're going to explore in the final project.

Final project

The key to learning to program well is learning to think like a machine. In other words, to be an effective programmer, a person needs to learn to think algorithmically, that is, in steps. One exercise that many programming instructors like to have their students do to master this skill is to have them take daily tasks and break them down into a series of steps. For the final project, we're not going to use a daily task but a task that one would normally see in the automation world.

A common task in automation would be to load parts onto a conveyor and start the machine. We're going to take this process and turn it into a series of steps that can later be converted into a program. The goal of this final project is not to produce pseudocode or a flowchart, as those concepts will be explored later. For now, all we need to do is produce a series of steps to accomplish our task. With that, let's look at loading the strategy we're going to use.

Strategy

To accomplish any task in programming, we need to look at the overarching problem and break that down into individual tasks. As such, we're going to break the problem down into two parts:

1. Loading the conveyor
2. Starting the machine

We can use these two steps as a roadmap to further break the problem down. So, the first thing we're going to do is tackle the conveyor steps. Before reading the steps, think about how you would load the conveyor. Write those steps down somewhere and compare them to the ones that follow. There are no right or wrong answers to this project. You can make the steps as detailed as you want; this is merely an exercise in algorithmic thinking. However, your algorithm should have at least three steps in it.

Conveyor

To load the machine, we will need to perform the following steps:

1. Bend down.
2. Wrap hands around the box.
3. Apply pressure around the box to grasp it.
4. Stand up.
5. Rotate to place the box over the conveyor belt.
6. Lower the box onto the belt.
7. Release grip.

For the operator to place a box on a conveyor belt, there could be roughly seven or more steps depending on how detailed you made your algorithm. With that part out of the way, we can now focus our attention on starting the machine. For this part, we're going to assume that there is a red light when the machine is off, and an E-Stop that needs to be released. After the E-Stop is released, the operator will need to press the blue reset button and then a green button. If everything is working correctly, a green light will come on.

Start machine

To start the machine, we could use the following algorithm:

1. Check whether the light is red.
2. Release E-Stop.
3. Press the blue button.

4.   Press the green button.

5.   Check to ensure the light is green.

Now, we have two algorithms that solve two different parts of the overall task we were assigned. Now, to complete the whole task, we can combine the two algorithms like so:

1.   Bend down.

2.   Wrap hands around the box.

3.   Apply pressure around the box to grasp it.

4.   Stand up.

5.   Rotate to place the box over the conveyor belt.

6.   Lower the box onto the belt.

7.   Release grip.

8.   Check whether the light is red.

9.   Release E-Stop.

10. Press the blue button.

11. Press the green button.

12. Check to ensure the light is green.

Now, the key takeaway from this exercise is to learn how to break big problems into smaller problems and think about simple tasks in a systematic way. Once those skills are mastered, you'll shine as a programmer.

Summary

In this chapter, we explored the basics of programming and programming languages. The goal of this chapter was to provide you with a gentle introduction to programming. We achieved this by exploring concepts such as programming languages, programming paradigms, program flow, the IEC 61131-3 standard, and programming logic.

The core takeaway from this chapter is to understand how programs work and flow at a high level, and to understand how to think like a machine. Thinking like a machine may be a completely alien topic now, but with practice, it'll become second nature. The best advice for those still struggling with algorithmic thinking is to pick a few more tasks, such as tying your shoes, combing, your hair, or taking a shower, and create little algorithms like the one in the final project. After you do that a few times, it will become second nature. Do not get overwhelmed by the size and scope of the task. If the task seems large, simply break it down.

Once you feel comfortable with algorithmic thinking, you are free to move on to the next chapter, which will cover the concept of computer memory. However, before you proceed, it is especially important that you feel comfortable with algorithms and the general gist of how a program works. Therefore, if you need to take a break and run through this chapter a few times, do so. The next chapter is going to be about unleashing the power of computer memory

Further reading

- IEC 61131-3

  `https://plcopen.org/iec-61131-3`

- Compiler versus interpreter

  `https://www.geeksforgeeks.org/compiler-vs-interpreter-2/`

- Machine instructions

  `https://www.geeksforgeeks.org/machine-instructions/`

- Programming paradigms

  `https://cs.lmu.edu/~ray/notes/paradigms/`

Questions

1. What is the difference between a compiler and an interpreter?

2. What is IEC 61131-3?

3. What is a machine instruction?

4. What is a programming paradigm?

5. Does IEC 61131-3 support OOP?

6. What is OOP?

7. Is IEC 61131-3-compliant code always portable between devices?

8. Write an algorithm for withdrawing $20 from an ATM.

9. What is a language translator?

10. What are the two types of language translators explored so far?

11. What are the languages that the IEC 61131-3 supports?

12. In what direction does a program flow?

13. What is syntax?

14. What are keywords?

Unleashing Computer Memory

The name of the game is memory. For a program to run smoothly, it will need adequate memory. **Memory** is a vital component of any programmable system. Computers have memory, microcontrollers have memory, and **programmable logic controllers** (**PLCs**) have memory. However, memory is a very complex topic that many automation engineers often take for granted. Typically, most automation programmers have an *if it can be uploaded to a device, it's fine* mentality. This mentality can get a person through their day-to-day activities, but it will hold them back in the long run, especially when more complex topics such as pointers need to be utilized.

Memory is a complex topic that can often baffle inexperienced developers. Memory has hardware aspects and software aspects that have to work in unison for a device to effectively function. The concepts that govern memory can often be very confusing for those who are not well versed in computer science, and as such their software and overall system will suffer for it.

This chapter is going to look at memory from a theoretical point of view. Looking at memory in this manner will show the complexity and intricacies of memory and how to effectively use memory in a PLC program. To do this, the following topics are going to be explored:

- What is computer/PLC memory?
- How does computer/PLC storage work?
- Volatile memory
- Non-volatile memory
- Memory addresses
- Common storage devices

To round out the chapter, we're going to explore what is stored in memory. Before we begin, it should be noted that many find the study of memory mechanics very boring. Unfortunately, understanding how memory works is of vital importance to effectively writing PLC software. If you fall into this category, think of this chapter as a quick necessary evil that'll only take a short time to read!

Technical requirements

This chapter is going to explore memory from a theoretical perspective; no software will be needed to follow along with this chapter.

What is memory?

Memory is often used as a catch-all term for data storage. That is, any type of storage is referred to as memory. However, grouping all types of storage together is not accurate. In actuality, there are two types of memory that are used in digital devices such as a computer or PLC.

Memory

Though it is easy to conflate storage with memory, in reality, memory is just temporary storage. This short-term memory is often referred to as **random-access memory** (**RAM**). RAM is used for a machine's immediate operations; for example, for running software such as the operating system or PLC code.

RAM allows the **central processing unit** (**CPU**) to quickly access program information from a storage device. As such, RAM can be thought of as temporary, high-speed memory. Fundamentally, RAM in itself is a microchip. However, for devices such as personal computers and even some PLCs, RAM will come packaged in a circuit board that contains multiple RAM chips. This whole device, colloquially called RAM, can then be installed in the motherboard of the device by means of a slot on the board. In terms of modern compact devices such as phones, tablets, and common PLCs, RAM is often baked into the motherboard and is placed close to the CPU chip. Common types of RAM are as follows:

- **Dynamic RAM (DRAM)**
- **Static RAM (SRAM)**
- **Double Data Rate SRAM (DDR SRAM)**
- **Double Data Rate Synchronous Dynamic RAM (DDR3 SDRAM)**
- **Rambus Dynamic RAM (RDRAM)**

The type of RAM in a device will vary depending on many factors such as price and application.

The other type of storage that is often commonly referred to as memory is called **storage**.

Storage

Storage is the term that most people use to refer to memory. As the name suggests, storage is used to permanently store data such as files, images, programs, and your operating system. In terms of PLCs, storage is often used to store things such as logged data and PLC programs. As logic suggests, the larger your storage volume, the more data you can store. This means that with a larger storage volume, more data can be stored, the larger your PLC program can be, and so on.

For most PLC programs, the standard storage volumes that come with CPU modules will suffice for most projects. However, in extreme cases, extra storage may be required. This is especially true for older PLCs that had very limited resources. Many PLCs have fixed storage, which means they cannot be expanded. However, high-end PLCs such as Beckhoff PLCs do have SIM card-like modules that are used to house the PLC's program, operating system, **human-machine interface** (**HMI**), and logged data.

Knowing whether your PLC's storage can be expanded or not is vital to the success of a project, especially if that project uses data logging. When large amounts of data are being constantly logged, the storage can fill up very quickly. Couple that with a hefty PLC program, HMI, and operating system, and the storage volume could be easily filled in a short amount of time.

Typically, if the project requires constant logging, especially if the logging is coming from multiple sensors, the data is usually stored on a separate machine. For applications such as power plants and large factories that are constantly collecting data, the PLC is networked into a larger **Supervisory Control and Data Acquisition** (**SCADA**) system or networked into some other software that connects to a database of some type. The database is usually either stored on a hefty server or, more recently, in the cloud where applicable. The PLC is usually rigged to act as a middleman to absorb and perform basic processing of the data and then pass the data along to the database via the chosen means.

In short, permanently storing data on the PLC itself is possible but is often not optimal. With the advancement of computing, and by extension PLCs, it is usually better to try to store long-term data on a device that isn't the PLC. Compared to the amount of storage in a modern server computer or the seemingly infinite storage capacity of the cloud, a PLC's internal memory will always be lacking, and should therefore be used as a last resort for the long term housing of data.

PLC memory isn't as robust as storing data on a server or the cloud. If one considers the environment that a PLC is often used in, one would realize that the conditions can be dirty and rugged and are often prone to extreme temperature fluctuations. This means that the life of a PLC is marred by conditions hostile to electronic components. If a PLC with fixed storage fails, all the data will be lost with the unit. On the other hand, even with a removable memory module, the data can easily become corrupted and lost due to failures with the PLC.. A common example of this occurring in higher-end PLCs, especially those that run an operating system such as Windows, is when the PLC goes into an emergency shutdown or rapidly loses power. If the system is writing to or reading from a file during the shutdown or power failure, the file could become corrupted. This means that whatever was in the file could be lost.

Overall, the only things that are wise to store on a PLC are the PLC program, HMI, other non-critical data, and so on. Not all PLCs can support database communication and, by extension, storage; however, many mid- to high-end PLCs support communication protocols that can be used to communicate with databases either on a server or in the cloud.

People often take the storage mechanism for granted. For many, the process of saving a file and having that file permanently accessible on the device is an act of magic. Most don't care how the data is stored as long as the data is there when they need it. For most, this is perfectly fine; however, to understand memory, it is important to understand how memory and storage work.

How computer/PLC memory and storage work

Growing up in the late 90s and early 2000s, it wasn't unusual to hear old-timers refer to memory as "*black magic.*" That is, they would simply write storage and memory off as something computers did just to work. However, as computer scientists, we need to be a little more diligent and understand how memory devices work. To begin the discussion we're going to explore how older, yet widely used, storage devices, **hard disk drives** (**HDDs**), work.

HDDs

HDDs once ruled the world of computer storage. HDDs could hold copious amounts of data and, for the most part, were reliable. HDDs were common in servers and personal computers for many years but eventually gave way to **solid-state drives** (**SSDs**).

HDDs could best be thought of as really large and reliable CDs. To summarize, the device works by magnetically encoding data to a disk, or as it is formally called, the **platter**. The platter is divided up into tiny subsections. When data is written to the patter, a section is either magnetized with the north pole facing up to represent a 1 or magnetized with the south pole facing up to represent a 0. When a program needs to be read off the disk, the head will go to where that program is stored on the platter and read the magnetized sections.

HDDs are excellent devices; however, they have drawbacks. One major drawback is that the disk can get worn out over time, the head will fail after a given amount of time, and they are very susceptible to magnetism. HDDs are also very slow compared to SSDs, noisy, large in physical size, consume more power, and are prone to mechanical failures.

For the data to be read off the disk, the disk has to be spun. The faster the disk is spun, the faster the drive is. However, there is a limit to how fast the disk can spin. Typically, a high-end commercially available HDD will spin at about 7,200 RPMs. SSDs, on the other hand, are typically 100 times faster, which means quicker boot times and program loading times.

SSDs

SSDs are the newer and faster iteration of storage devices. In many ways, SSDs can best be thought of as flash drives on steroids. However, unlike HDDs, SSDs do not have any moving parts, which is why they are significantly faster than HDDs. The main difference between the two types of devices stems from the storage mechanism itself. Where a traditional HDD uses a magnetized disk to store data, an SSD will usually use some form of flash memory to store data.

Flash memory means that data is electrically written to a series of chips. Ultimately, this means that the data can be quietly written and read from the devices. This also means that mechanical failures such as a stuck head or worn-out platter won't spell the end of the hard drive. Typically, a high-end SSD will have the following advantages over an HDD:

- Faster read/write speeds
- Less power consumption
- Better durability
- Permanent deletion of data
- No noise

These are just a few attributes that demonstrate the superiority of SSDs over HDDs, but there are many more.

Though superior in many respects, SSDs do have a few drawbacks. Major drawbacks include the following:

- **Shorter lifespans**: Lower-quality SSDs that use NAND for flash memory often suffer from a limited number of read/write cycles.
- **Price**: SSDs typically cost significantly more than HDDs. The higher the storage capacity of an SSD, the more it will typically cost.
- **Storage capacity**: Though SSDs are quickly catching up in terms of storage capacity, they are still limited compared to HDDs. In terms of pure storage potential, HDDs are still slightly superior.

There are also other disadvantages of SSDs. Traditionally, SSDs offer significant advantages over HDDs. Most of the disadvantages, including the ones listed previously, are usually found in lower-end SSDs.

HDDs are still commonly used in manufacturing equipment. It is very common to have to troubleshoot devices that use HDDs, and many times those HDDs are the culprit. When applicable, it is wise to replace HDDs with SSDs when the HDD device fails.

In terms of PLC storage, most modern PLCs use some type of solid-state storage for both the program and any data that is stored on the device. Generally, solid-state storage is more durable than HDD storage, which means it will perform better in the rugged environment in which PLCs usually operate. Now that the different types of storage have been thoroughly explored, we can move on to exploring the differences between the concepts of **volatile** and **non-volatile memory**.

Volatile versus non-volatile memory

To keep things consistent with everyday speech, we're going to continue to use the term "memory" for any device that can store data. Though we did make a distinction in the prior section, from here on out, the term "memory" will refer to either a storage device or a device such as RAM. In terms of application, there is a major difference between volatile memory and storage. Essentially, the difference between the two types of memory boils down to whether or not data is permanently stored after a machine is power cycled. To begin the discussion, let's explore volatile memory.

Volatile memory

In a technical sense, the best way to describe volatile memory is as non-persistent memory. Non-persistent memory is a fancy way of saying that the data stored is lost when the chips lose power. A common example of volatile memory is RAM. When a program loads, it gets dumped into RAM, where it is then executed. The program will track changes as long as there is constant power to the device. However, once the power has been removed, all these changes will be permanently lost.

Typically, changes made to data in the PLC's memory, called **variables**, are volatile. This means that if an operator enters the number of parts a machine needs to make, or inputs a timer to keep track of how long a process has been running, these values will be lost if the PLC is shut off for any reason. Typically, this won't matter much; however, for long-term storage of the PLC program itself, variable values that must persist during a shutdown phase, and more non-volatile memory must be used.

Non-volatile memory

Non-volatile memory is often referred to as **persistent memory**. Persistent or non-volatile memory is memory that can keep data stored even when the device is not powered. Traditionally, a non-volatile memory device would be something such as an SSD or HDD. Where RAM is an example of volatile memory, **read-only memory (ROM)** would be an example of non-volatile memory.

Non-volatile memory is of essential importance to PLC programming. All PLC programs and collected data have to be stored in non-volatile memory. Obviously, a PLC programmer would want their program to persist even when the PLC is powered down. However, persistent memory is used for much more than just storing PLC programs. As was stated before, it can be used to store data long-term as well.

Quite often, PLC programmers will find themselves needing to retain settings even after a power cycle. One such application is calibration data. Calibrating machines can often be a daunting task that is very detailed and can take a lot of time. If all the calibration data was lost when the PLC was powered down, the company could lose thousands of dollars to get the machine operational again. For data such as this, one has to store the data in persistent memory.

Luckily, many higher-end PLCs, especially those that follow the IEC 61131-3 standard, have what are called **persistent variables**. For the most part, a persistent variable is a place in memory that is non-volatile. For data such as calibration data or configuration settings, the programmer can opt to store this data as a persistent variable for permanent storage.

Memory, in general, is a very powerful computer science concept. There are many ins and outs to how memory works and organizes itself.

Memory addresses

Up until this point, we have explored the types of memory and memory devices. However, what we have not explored is how memory works. As such, this section is going to be dedicated to exploring how memory works.

How memory works

Conceptually, memory can be visually represented by *Figure 4.1*:

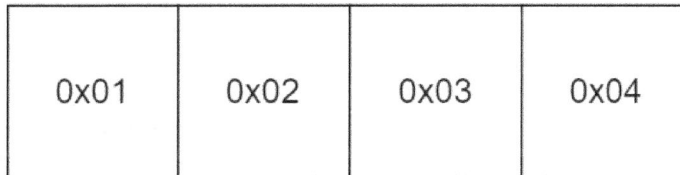

0x01	0x02	0x03	0x04

Figure 4.1 – Computer memory representation

Computer memory can best be thought of as tiny sections that compose the overall memory system. In the case of *Figure 4.1*, each memory square is a **memory block**. That is, each square will hold a piece of data such as the number of parts to make or the state of the machine.

Each memory block has an address that is used by the PLC to organize and keep track of the memory locations. In the case of *Figure 4.1*, the address is the alphanumeric label in the diagram. When data needs to be retrieved, the computer or PLC will invoke the memory address and retrieve the value. On the flip side of that, if the machine needs to insert or change the value of the data, it will again invoke the memory address and inject the new value into it.

Computer memory – an analogy

The technical explanation for computer memory can be a bit jarring and confusing for the inexperienced. However, the best way to think of computer memory is as an apartment complex. A typical apartment complex is a building subdivided into separate units. In the case of memory, the building would be the storage medium such as the platter on an HDD, and the apartment units would be the memory labeled memory blocks.

Also, similar to the way apartments have unit numbers, the memory cell has a unit number as well (that alphanumeric string). When a new resident moves into an apartment unit, they will be assigned a unit number. When they move out, the apartment complex owner will be aware that the resident is leaving. This is almost exactly how computer memory works. When something needs to be written to a memory block, the device will select and open the memory block and let the data move into that unit. If the data ever needs to be retrieved, it will go to that address and pull the data out so it can be used.

Memory addresses are very powerful concepts and can be used to great effect in PLC programming. Most PLC programming environments will allow a developer to directly access a memory address and modify the data directly. This is an advanced concept called pointers. Using pointers to directly access data in memory comes with great risk, as it can create what are called **memory leaks**, which can harm the efficiency of the PLC program or even crash it.

Typically, the memory address that a machine will use will not be readable by a human, nor will it provide much context as to what it is used for. To efficiently use memory, a developer will need to use a variable, which can be thought of as a human-readable alias, for the address. Variables or tags will be explored in more detail in later chapters. For now, we're going to shift our attention to storage devices.

Common storage devices

It is hard to exist in the 21st century without knowledge of common storage devices. However, for the sake of being thorough, it is important to cover these concepts and how they can be used in an automation environment. With that, the first device we're going to explore is a USB flash drive.

USB drives

Probably the most common storage medium in use are **USB storage devices**. These are small, portable storage devices that can be used on any computer. To use a USB drive, all one has to do is plug it into a USB port and let the driver software automatically install. These devices can hold copious amounts of data with some high-end models capable of storing over 16 TB of data!

For automation applications, USB drives are mostly used to move data from device to device. Obviously, USB devices are mostly used with higher-end PLCs that have USB ports and utilize an operating system that can support USB drives, such as Beckhoff PLCs, which run Windows. These devices can also be used with automation systems that live on computers, such as SCADA systems. These devices are used to transport data such as files, logs, and even small programs from machine to machine. USB drives are not the only storage devices that utilize USB ports. Another device that utilizes USB ports is a USB drive's big brother, an external hard drive.

External hard-drives

Another USB storage device is an **external hard-drive**. In terms of modern technology, an external hard-drive can have significantly more storage than a USB drive. External hard-drives also work in a similar way to USB drives. Both types of devices require a USB port and a compatible operating system to operate. However, there are a few advantages to external hard-drives.

Compared to a USB drive, an external drive is advantageous in that it is more durable than a USB drive and it is a little faster. Another feature that some consider an advantage is that external drives are physically larger than USB drives, and are, therefore, harder to lose. This makes them especially handy in automation applications, as they can be integrated directly into the system. In some cases, engineers will create brackets for the external drives and mount them on a DIN rail where they can quickly be removed from the bracket.

Regarding automation engineering, these devices are meant for added long-term storage. For example, it is common to use these drives with systems that incorporate cameras for image detection. For systems such as these, it is often important to keep the images long-term, so they are stored on the removable drive until they can either be removed or backed up somewhere else. As such, these devices make an excellent option for devices that need extra long-term storage and are not hooked up to the cloud or other external computer systems.

USB-based devices are not the only type of external storage that can be used with PLCs. Another common device is one that is used with a smartphone or tablet. That device is an SD card.

SD cards

Unlike USB devices, an **SD card** does not need a USB drive to plug into. However, an SD card will need an SD card slot on the PLC, and the PLC will have to be able to support the device. SD cards are not meant for the same type of storage that USB drives and external hard-drives are used for. Where USB devices are mostly used to store data such as pictures or transport data, SD cards are mostly used to expand the storage capacity of the device. As such, when configured properly, the device will allow for larger and more complex programs.

SD cards come in two types. Nowadays, the more common SD card is called a **microSD card**. These cards are smaller but can store large amounts of data like USB drives. The other type of SD card is a regular SD card, which is just physically larger than a microSD card. When using an SD card, it is important to know whether the device supports a micro or regular SD card. Typically, an SD card adapter can be used if all that is on hand is a microSD card, and the device is set up to only support a full-sized card. The microSD card will plug into the adapter, which will allow it to be used in ports that are meant for full-sized cards.

Cloud storage

Though not technically a device, the newest type of storage is **cloud storage**. As has been briefly mentioned in the past, cloud storage is storing data in third-party storage devices. This is usually accomplished by connecting your application to the internet and routing data to an offsite storage device that exists in the cloud. The cloud provider will offer storage for the data and regular backups. This means that the end users will never have to worry about losing their data. This is a very reliable and safe option for permanent storage, but it can be costly and does require an internet connection. If your application is air-gapped, that is, not connected to the outside internet, or the end user does not want to pay for the storage, this may not be an option.

Obsolete storage devices

Technology that is used in automation can be very old and outdated. Systems can be in place with relatively minor changes for 20+ years. As such, as an automation engineer, it is not uncommon to see very antiquated equipment, and storage devices are no different. To be successful as an automation engineer, it is important to have a basic understanding of what the older storage methods were, just in case you see one in the future.

CD and DVD-ROMs

CD and DVD-ROMs are not necessarily old technology used for long-term storage, but they are falling out of favor. Chances are, you've probably seen and used some type of CD and/or DVD-ROM in the past. You may have listened to your favorite band's latest album on CD or watched your favorite movie on DVD. However, those still somewhat common CD and DVD-ROMs are being used a lot less frequently. For storage, CD and DVD-ROMs have been taken over by external hard drives, USB thumb drives, or cloud storage.

Floppy drives

If you were born after the year 2000, there is a chance you've never seen a floppy drive before. A floppy drive is a 3.5-inch rectangular piece of plastic with a magnetic strip in it that is used to store data. These storage devices ceased production around 2010 and are incredibly rare in today's world. However, due to the age of some automation systems, it is not impossible to come across these devices.

These devices have moving parts in them that can easily wear out. This can cause a bit of a pickle since floppy disks are no longer produced. Therefore, if you come across a system that requires a new floppy disk, the system is automatically going to have to be upgraded when it fails. If the drive was used to store any software, the software will have to be rewritten and the storage mechanism will need to be changed to something more modern, such as an external drive.

Typically, floppy drives are used as a transfer mechanism between old PLC-based systems and old computers. Seeing a floppy disk in a PLC-based machine is a key indicator that the system is in desperate need of an upgrade. Usually, these systems will require new storage mechanisms, PLCs, and other modern hardware.

Summary

This chapter has been a crash course on computer memory with an emphasis on storage. Topics explored include different types of memory and storage, storage devices, and the inner workings of memory. There is much to know about how memory works and whole books are dedicated to it. However, by this point, you should have a good grasp of how memory works and what it's used for. Now that memory has been explored, in the next chapter, we will explore another very important concept in computer science: designing a program.

Questions

1. How does an SSD work?

2. What is a memory address?

3. What is an example of an obsolete storage device?

4. Name two modern storage devices.

5. What is cloud storage?

6. What are two drawbacks to cloud storage?

7. What is a memory block?

8. What is volatile memory?

9. What is non-volatile memory?

10. What kind of memory is ROM?

11. What kind of memory is RAM?

12. What is storage?

13. What does RAM stand for?

14. What does ROM stand for?

Further reading

- *How A Hard Drive Works*

  ```
  https://cs.stanford.edu/people/nick/how-hard-drive-
  works/#:~:text=The%20hard%20drive%20contains%20a,the%20stored%20
  0's%20and%201's
  ```

- *SSD Advantages and Disadvantages*

  ```
  http://www.laptoppricelist.in/kb/ssd-advantages-disadvantages
  ```

- *Volatile Memory vs. Nonvolatile Memory: What's the Difference?*

  ```
  https://www.trentonsystems.com/blog/volatile-vs-nonvolatile-
  memory#:~:text=At%20a%20high%20level%2C%20the,after%20the%20
  system%20shuts%20off.
  ```

- *What is cloud storage?*

  ```
  https://aws.amazon.com/what-is/cloud-storage/
  ```

5

Designing Programs – Unleashing Pseudocode and Flowcharts

Believe it or not, a quality program requires a design. Not designing a program is a lot like driving to work without knowing where you're going. This is something that is often overlooked by inexperienced programmers and, especially, PLC programmers. To ensure you have a quality, efficient program on your hands, you need to design the algorithm before you even think about touching a keyboard.

Much like an electrical engineer would never start building a circuit without creating a diagram, a programmer should never start writing a program before at least a rough design is implemented. A program design is a lot like a roadmap for the programmer. Similar to the way a map will save a driver many hours of driving around aimlessly, a program design will save a programmer countless hours of trial and error, as well as provide documentation that other developers can follow should they also contribute to the project or inherit the project in the future.

Designing a program is multifaceted, the same way designing an electrical or mechanical system is. A program is usually designed at a high level, and then each component is subsequently designed. There are a lot of different design techniques out there, but for this book, we are going to look at two techniques called pseudocode and flowcharts.

To understand how to design a program, we're going to look at the following concepts:

- What are pseudocode programs and flowcharts?
- Why use pseudocode and flowcharts in PLC programming?
- What is the difference between flowcharts and pseudocode?
- Tools needed to implement pseudocode and flowcharts

- Techniques for using pseudocode and flowcharts
- Example exercises that use both techniques

Finally, to round out the chapter, we are going to design a robot startup program. For the final project, we're going to build the program in pseudocode and with a flowchart to drive in the concepts.

Technical requirements

To follow along with this chapter, access to `draw.io` will be needed for flowcharting, and some type of text editor will be needed for pseudocode. For this chapter, any text editor can be used. Recommended text editors are as follows:

- Notepad
- Notepad++
- Microsoft Word

What are pseudocode programs and flowcharts?

The first step in developing a very robust and long-lasting program is understanding design methodologies. Depending on the design level – that is, if you're working on a small component of the system or the overarching architecture – there's a design methodology that will suit your needs. For this book, we're going to be concerned with the two most common methodologies that are employed by software developers of all types: pseudocode and flowcharts. To begin our design discussion, we need to first look at pseudocode.

Pseudocode

One of the easiest design tools a programmer has at their disposal is **pseudocode**. Pseudocode is an extremely simple concept that can help programmers work out their programs in everyday language. However, the sheer simplicity of the technique will often stump inexperienced programmers.

What is pseudocode?

Pseudocode is not a programming language, nor is it meant to produce a working product. Instead, pseudocode is a design tool that is meant to help developers express their would-be programming logic in a way where they are not constrained by things such as syntax or programming structure. In other words, pseudocode is a way for programmers to work through their programs' logic using everyday language.

Developing programming logic with pseudocode is unique compared to many other design tools. Pseudocode is not defined by any rules. There is no syntax, no grammar, symbols, or anything else to memorize or learn. Pseudocode can be written using the developer's own natural language and with their natural way of speaking. This means the only real "rule" is that the developer, along with anyone else, should be able to understand pseudocode when they revisit it in the future.

Now, there are no rules that directly govern pseudocode; that is, there is no true standard to follow. However, some organizations do implement rules on how developers should structure their pseudocode projects. These instances are usually relegated to places that require heavy documentation or in academia. Many undergraduate programming programs will usually have a format for students to follow; however, as stated before, once a student graduates, the only time pseudocode format will matter is if they are employed with an organization that has very strict documentation standards.

What does pseudocode look like?

To understand what pseudocode is, the best thing to do is to look at an example. To demonstrate pseudocode, let's explore a simple program that can take user input, calculate the area of a circle, and finally display the calculated area:

```
radius = input
area = 3.14 * radius^2
display(area)
```

As can be seen, this example uses a simple intuitive syntax. To represent an input command, we simply used a variable assigned to the word `input`. The next line takes the proverbial input on the first line and performs a calculation with it. Finally, the last line represents displaying the computed area on the screen.

The key takeaway from the example is that pseudocode can be written any way you want it to be. However, you want it to make sense to yourself and to others. For example, each line in the example is clearly defined and clearly states what its purpose is. This means that all one must do to turn this into a real program is to take the lines in the example and convert them to the desired programming language.

> **Note**
> Pseudocode is not meant to produce a working program; it is just the symbolic representation of the general logic for a working program.

Overall, pseudocode is a very common design technique to help programmers work out their logic. This technique is especially useful for inexperienced developers looking to iron out their logic before they are bogged down with things such as syntax and the programming language's grammar. Though it is a very popular and useful tool, it is not the only design technique that can help developers work through their logic. Another very popular design technique is a technique called flowcharting.

Flowcharting

Many people prefer graphical representation to written tasks. Naturally, software developers will often opt for a graphical tool as well. One way to graphically represent programs is with flowcharts.

Flowcharts are very prevalent in automation engineering. Flowcharts are so prevalent in automation that there is even a programming language governed by the *IEC 61131-3* standard that uses flowcharts to write programs. There are also programming systems such as Flowgorithm that can be used to write actual programs, albeit these systems are usually used as teaching tools in academia or for beginners learning to program.

What are flowcharts?

As stated before, flowcharts are graphical representations of a process. Since programs are processes, they are excellent tools that graphically represent the flow of a program. Unlike pseudocode, there are rules that govern how flowcharts should be drawn, mainly the meaning of the symbols. As such, when learning to flowchart out a program or process, it is first necessary to understand flowcharting symbols and what they mean.

Flowchart symbols

Basic flowchart symbols can be found in *Figure 5.1*:

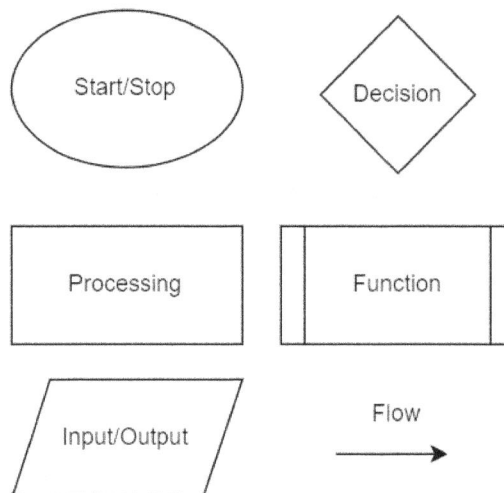

Figure 5.1 – Basic flowchart symbols

Each of these symbols is kind of like keywords in a regular programming language. The following is a breakdown of what each symbol does:

- **Start/Stop**: As the name suggests, this symbol represents the beginning and end of the algorithm. It is not uncommon for designers to omit these symbols. Technically, though, the very first symbol should be a circle with the word *Start* in it, and the last symbol should be the very last symbol with the word *End*. Note that the words *Start* or *End* may be replaced with different words depending on the system or standard that is being used.

- **Decision**: This symbol represents different paths that a program can take. In other words, these are conditional statements equivalent to an `if` statement in a general-purpose programming language. Actual flow control will be explored in detail in *Chapter 11*. For now, just remember that a decision or flow control statement simply means that it is a different path a program can take under a certain condition.

- **Processing**: Processing symbols are used to represent individual steps in a program that "*do things*." For example, a processing command can represent adding numbers, processing a string, declaring a variable/tag, or anything along those lines. In other words, processing is just a representation of a generic operation in the program.

- **Function**: Functions are an advanced topic. A function is a block of code that performs a task. At first glance, this may seem like a processing symbol or keywords; however, a function is a group of operations that perform a specific task. For example, all languages support the basic operations of addition, subtraction, multiplication, and division. However, many languages don't have built-in support for calculating the area of a circle. A function would be a group of commands that has all the necessary logic to compute the area of a circle. More on built-in functions will be explored in *Chapter 10*.

- **Inputs/outputs**: As the name suggests, inputs and outputs represent inputs from the user and outputs from the program respectively. An example of an input would be an operator inputting the number of parts that need to be created during a production run, while an output would be the current number of parts that have thus far been produced.

- **Flow**: The arrow represents the flow of a program. For the electrically inclined, these symbols can be thought of as wires in a wiring diagram. In short, flow symbols are used to connect other symbols together in a sequential order. The only real rule to remember with flow symbols is that the arrow at the end of the line will point to the next step in the process.

It should be noted that depending on the standard set by the organization you're performing work for or the software you're using to generate a flowchart, the symbols may vary a little, as well as the number of available symbols that can be used. However, the general gist of the symbols, along with the logic that governs them, will usually be consistent.

What do flowcharts look like?

Much like pseudocode, to fully understand flowcharts, it is important to understand what they look like. So, to get an idea of what a flowchart looks like, let's take the circle area program in pseudocode form discussed earlier and convert it to a flowchart:

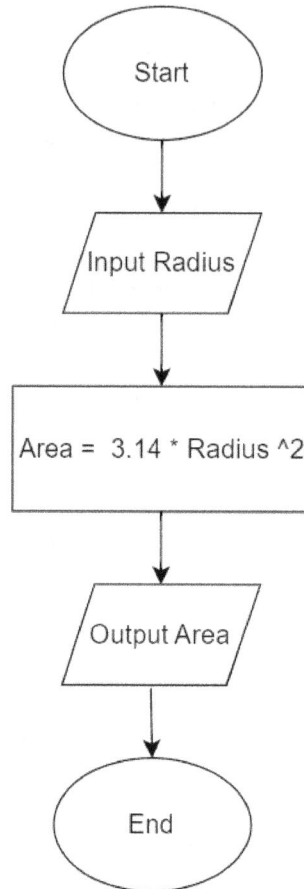

Figure 5.2 – Flowchart for the circle area program

As can be seen, the flowchart is very similar to the pseudocode. The only real difference is that the flowchart used symbols, but the general flow of the program is the same. In the flowchart, the following was depicted:

- The initial oval represents the program's start.

- The parallelogram represents a user input command – in this case, the radius.

- The next rectangle represents the area computation.
- The second parallelogram represents the output of the computed value. That is something akin to a `print` statement.
- The final oval represents the end of the program.

Flowcharts are usually drawn in a top-to-bottom manner, but this is not a steadfast rule. The main thing to pay attention to in a flowchart is the direction in which the arrow is pointed. Notice that they are pointed toward the bottom of the page, which in this case is the next step in the process or program. It is very important to ensure that the arrow is always oriented the correct way as it can easily cause confusion when an engineer tries to implement the process in real life.

Understanding pseudocode code and flowcharts is only half the battle in designing a program. The next vital step in designing a program is understanding why one would opt to use these techniques.

Why use pseudocode and flowcharts in PLC programming?

So, why use pseudocode or flowcharts for designing a program? To begin this discussion, let's first explore why we would use pseudocode.

Why use pseudocode?

As we established in the previous section, pseudocode is a design technique that allows developers to express their logic in everyday language so that they don't get bogged down with the programming language's syntax or grammar. This is a grand explanation of what pseudocode code is, but it fails to address why we should use it:

- **Design tool**: Above all else, pseudocode is a way for developers to iron out logic and pitfalls in an algorithm or processes(es) without having to worry about programming grammar. In other words, pseudocode is a way of relaying the logic of an algorithm from one developer to another or their future self in an easy-to-understand format.
- **Porting tool**: Since pseudocode is language-neutral, it is commonly used to take existing algorithms in other languages and express them in a language-neutral way. This makes porting code very easy as only the developer who is writing out the pseudocode needs to understand the original programming language.
- **Documentation**: Pseudocode can also be considered rough documentation. It is not uncommon to include a conceptualized draft of a high-level algorithm or process in the documentation. This includes when the team hasn't fully fleshed out the design of the system and languages that are going to be used are not fully agreed upon yet. For example, the team may not know if they want to use C# or Delphi for the **human-machine interface** (**HMI**) yet, but they do have an idea of how the user should enter data into the HMI. In cases such as these, pseudocode is an excellent way of expressing how the algorithm should work while keeping it language-neutral.

- **Interviews**: It is not uncommon for would-be employers to "test" potential employees on their technical prowess. Whether or not this is a quality practice is up to debate; however, if this practice is used, a quality employer will usually request that the candidate use pseudocode. Quality employers will typically opt for the candidate to use pseudocode or flowcharting over writing working code to alleviate some concerns by would-be employees who fear their *test code* will be used in a real-world application without them being compensated.

These are just a few ways to use pseudocode. Pseudocode is just a tool; as such, the only limiting factor in how it can be used is the developer's imagination. Much like pseudocode, flowcharting can be used in a very similar fashion.

Why use flowcharting?

Flowcharting is used in many different engineering fields for many different things. One other profession where flowcharts are used quite often is in the field of business, where business processes can be graphically represented. Regardless of the field, flowcharts are used to do the following:

- Design a process
- Identify and remove redundancies
- Pinpoint pinch points in a process
- Optimize a process
- Troubleshoot issues in a process
- Document processes
- Document brainstorming sessions
- Help interview candidates

Typically, flowcharting is not used to describe the steps in a program; instead, it is used to describe the overall system. For example, if an automation system has multiple steps in a process, flowcharting can be used to describe how all those processes interact with each other. In other words, a flowchart typically wouldn't be used to depict a program; instead, it would be used to depict something akin to a plant process.

Brainstorming

Brainstorming is one of the biggest use cases for flowcharts. Due to the high-level nature that is often associated with flowcharts, they make great tools to create theoretical mockups for an automation system. Typically, during design meetings, flowcharts are heavily employed to depict how the data will flow in a system and how each software component will interact with one another.

When to use one over the other?

As a developer, when should you use one technique over the other? This is a logical question and one that we have touched on a bit. However, we can summarize when to use the two techniques with the following:

- **Pseudocode**: A developer should use pseudocode to represent the source code. Pseudocode should be used to work out a rough outline of a program using everyday language. Usually, pseudocode is used to generate rough documentation for a program's logic, as a tool for porting logic from one programming language to another, or as a design tool. Pseudocode can also be used for brainstorming; however, pseudocode should be used to brainstorm lower-level systems such as the actual program components that are going to be implemented on a PLC. Overall, pseudocode is used for low-level system tasks.
- **Flowcharts**: Flowcharts are graphical design tools that work best for depicting high-level processes. Flowcharts can be used to represent low-level source code, but they are best suited to depict processes such as a whole plant process. In terms of design, flowcharts are often used in brainstorming sessions to depict the data flow for the system.

That said, both flowcharts and pseudocode are tools. A developer can opt for whichever tool works best for them. If a flowchart works better for a given task, then it is perfectly acceptable to use it. On the other hand, if pseudocode works better for a task, then it is perfectly acceptable to use that technique.

Hopefully, it was demonstrated that both pseudocode and flowcharting are excellent and easy-to-use design tools. However, how does one go about creating either a pseudocode program or a flowchart? The answer is surprisingly simple and will be explored next!

Tools needed for flowcharts and pseudocode

A very common question for students and inexperienced engineers regards which programs they should use to write pseudocodes or draw flowcharts. These are very common and very logical questions, and the simplicity of the answer often baffles the uninitiated. As such, the following section will explore some common tools that can be used to design a program.

Pseudocode tools

As can be seen from the *Technical requirements* section, pretty much any text editor can be used to write pseudocode. Typically, engineers and students will use something simple such as Notepad in Windows to write pseudocode. For the most part, an engineer can use anything they want to jot out pseudocode. Some will even opt to use a pen and paper, and in interviews, it is not uncommon for would-be employers to write out pseudocode on a whiteboard. The following are some general editors that are often used to write pseudocode:

- Notepad
- Notepad++

- Word
- LibreOffice
- Any **integrated development environment (IDE)**
- Vim
- Nano
- Text editor
- Pen and paper
- Whiteboards

As can be seen, a developer can use their favorite text editor to write pseudocode.

Flowchart tools

Producing flowcharts is a bit more complex than producing pseudocode. As was seen in the flowchart example, flowcharts require specific symbols to make sense. This means that special software will be needed to draw flowcharts. Any program that can draw the symbols or similar symbols will suffice. However, common programs that can be used to draw flowcharts are as follows:

- `draw.io`
- Visio
- Lucidchart
- SmartDraw
- Flowgorithm
- SFC PLC programming interface
- Pen and paper
- Whiteboard

This is not a complete list; there are many ways to draw a flowchart. Each program will put its own twist on how flowcharts are drawn. This means that the symbol types and number of available symbols might vary between the rendering software. So, it is important to keep in mind that if an engineer is accustomed to using one type of rendering software and they are required to switch to another system, there may be a bit of a learning curve to get up and running with the new program.

Flowchart programming systems

Now, the important thing to remember is that systems such as SFC and Flowgorithm are designed to produce working software. This means that though you can draw a flowchart with these systems, they are ultimately meant to produce working software, not just draw it. In other words, these interactive systems are *NOT* design tools and should only be used in a pinch!

These interactive systems can be very difficult to design with. Since they are actual programming languages, they have syntax and grammar rules that must be followed. This means that these systems may throw errors, change colors, or even prevent the engineer from continuing their drawings until certain syntactical errors are fixed. Ultimately, using an interactive system can very easily divert attention from working out a design to writing a working program, which is not what an engineer wants to do at this stage.

Of all these methods, it can be argued that whiteboarding or using pen and paper is the most common way to draw a flowchart or write a pseudocode program. As such, the next section is going to explore why this is so prevalent and why it is important.

Whiteboarding

Ironically, the most used medium for writing pseudocode or drawing flowcharts is a whiteboard. This may seem oxymoronic in the modern, hi-tech landscape. However, the following will dive into why drawing flowcharts and writing pseudocode by hand is so prevalent and why it is important.

Handwritten pseudocode

For organizations that do not require pseudocode, it is very common to write code by hand when necessary. Typically, when people write things down by hand, they grasp the information better. Following this logic, it can be argued that it is sometimes more effective to write pseudocode by hand.

High-level pseudocode is also heavily used in brainstorming sessions. This ultimately means that writing pseudocode by hand is very common for developers. It is recommended that inexperienced programming professionals and students should get into the habit of writing pseudocode by hand to get a feel for it. Writing pseudocode by hand can be a different experience than writing it on a computer for some.

Flowcharts – Whiteboarding

As with pseudocode, the most common way to draw a flowchart is by hand. Flowcharts are often used to depict high-level systems. In other words, they depict how individual software systems interact with each other. For example, it is common to depict the data flow from a PLC sensor all the way to a database via flowcharts. Because flowcharting is commonly used in brainstorming sessions to depict high-level mockups of theoretical systems that are under design, the most common way to draw is with a dry-erase marker on a whiteboard. Once a rough architecture of the system is agreed upon, actual rendering software such as Visio or draw.io will be used to make more permanent drawings.

Technical interviews

As has been mentioned in the previous sections, it is very common to be asked to either draw a flowchart or write a pseudocode program for a would-be job. Flowcharting and pseudocode exercises are language-neutral ways to evaluate a candidate based on their design or coding skills. Typically, the design technique used will vary depending on the job the candidate is going for.

For coding jobs – that is, jobs for writing PLC or HMI code – the candidate will typically be asked to write pseudocode to evaluate their coding and thought process. During a technical interview, candidates are sometimes asked to pick up a marker and start pounding out a program. On the other hand, flowcharting is used more in system design interviews. For example, if a candidate is going for a design job that requires designing a process or the digital side of a machine, they will usually either opt for designing the mockup using a flowchart or be flat out required to.

Why are interviews and using these design techniques so important to this conversation? The answer is surprisingly important. As stated before, it is very common for a candidate to be required to do a system mockup before a company will offer a candidate a job. Typically, for companies that require this, they will usually design the system by hand on a whiteboard or paper. However, for many engineers, drawing or writing by hand in a stressful situation can often be very challenging and requires a bit of practice, especially to draw a coherent flowchart where symbolism matters. Therefore, it is very important to get used to working out designs both by hand and with some type of computer program.

Now, learning how to properly design by hand is a skill; however, it is only half the battle. The other half is learning to think like a machine to architect a well-crafted program, and the key to that is practice. In the following section, we're going to work through a few problems to get into a machine mindset so that we can create quality algorithms and systems.

Design exercises

The biggest challenge that many programming students have with learning to program is learning to think like a machine. The biggest hurdle many programming students have with learning to think like a machine is learning to think in sequential steps. As with anything else in life, the key to learning to think algorithmically is practice, practice, practice!

The quadradic equation

To begin, let's assume that we must design a program that can take three inputs and must output the results for the following equation:

$$x = \frac{-b \pm \sqrt{b^2 - 4ac}}{2a}$$

Before continuing, try taking a moment to lay out the steps of how the program should work and the designs.

Design logic

The first step in designing this program is working out the steps that the program would need to follow. Since this is a math equation, we can apply rules that are commonly used in math to write the program. Therefore, the steps that can be used are as follows:

1. Get the inputs for a, b, and c.

2. Compute -b in the equation.

3. Compute the square of b and subtract the product of 4, a, and c from it.

4. Add the output from *step 3* to the output of *step 2*.

5. Subtract the output from *step 3* to the output of *step 2*.

6. Divide *step 4* by the product of 2 and a.

7. Divide *step 5* by the product of 2 and a.

8. Assign the output from *step 6* to a variable.

9. Assign the output from *step 7* to a variable.

Pseudocode

The pseudocode for this program can be easily streamlined and could resemble the following:

```
a = input(A)
b = input(B)
c = input(C)
negativeX = [(-1 * b) - sqr([b^2] - [4*a*c]) / (2*a)]
positiveX = [(-1 * b) + sqr([b^2] - [4*a*c]) / (2*a)]
print root1 is negativeX
print root2 is positiveX
```

Now, this is one rendering of how the pseudocode can look. This design code is not meant to be compiled and run as is; all that matters is the logic that the pseudocode is meant to portray. This means you can use any verbiage you want when you're working through this exercise, and you can use as much detail as you want.

Flowchart

The following figure shows a quadratic equation flowchart:

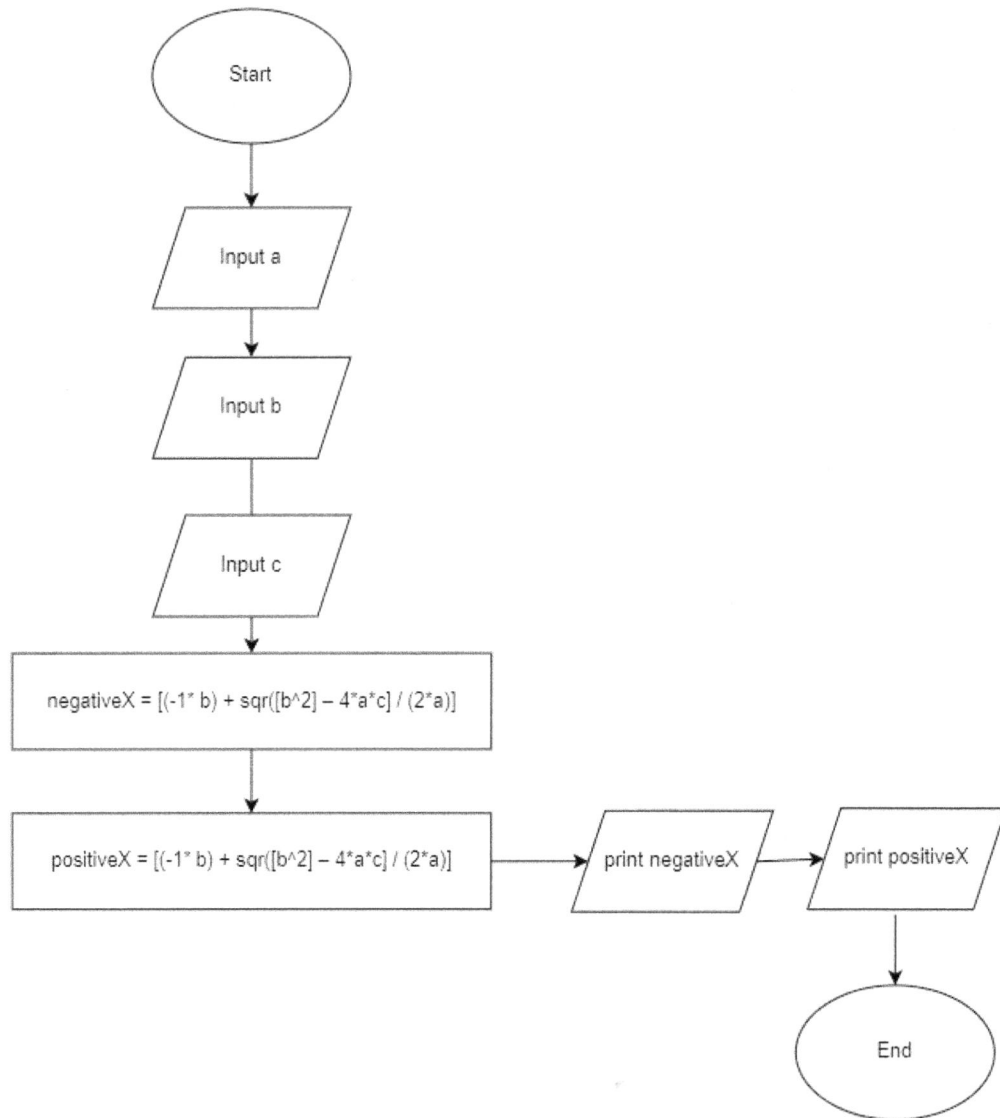

Figure 5.3 – Quadradic equation flowchart

Now that we've worked through a basic example, let's move on to something a little more complex. Let's work on a program with some flow control in it.

The beer program

In the United States, a person must be 21 or older to buy alcoholic beverages such as beer. So, for this exercise, let's design a program that can determine if a person is eligible to buy beer in the United States. As in the last example, take a moment to try to work out this problem before you continue.

Design logic

For this project, the following logic can be used for the system:

- Input the customer's age.

- Should the customer be at least 21 years old, let them buy beer.

- Should the customer be under 21 years old, throw a shoe at them.

These should be the general steps for the program. At first glance, this program may seem less complex than the quadratic equation program. However, the logic for this program will be less linear as the program can take two different paths. In other words, the algorithm will require a control statement. Control statements will be fleshed out more in *Chapter 11*, but for now, just think of a control statement as a branch in a program. With that, fleshing out the program a little more will yield the following pseudocode.

Pseudocode for beer program

In programming, flow control is usually defined with a command called if. In pseudocode, it is common to use the word if to denote its programming language counterpart. Therefore, the pseudocode could look like the following:

```
age = input(person's age)
if age greater than or equal to 21
    print "let them buy beer"
if age is less than 21
    print "throw a shoe at them"
```

Normally, a programmer would use an else or else if statement to denote the second condition; that is, if the customer is under the age of 21. However, that is out of the scope of this chapter. For more on else and else if statements, see *Chapter 11*. However, for now, we're going to simply use a second if statement to process the second condition. With the pseudocode now in place, let's tackle the flowchart.

Flowchart

For this project, the flowchart may not be one-to-one with the pseudocode as in the past examples:

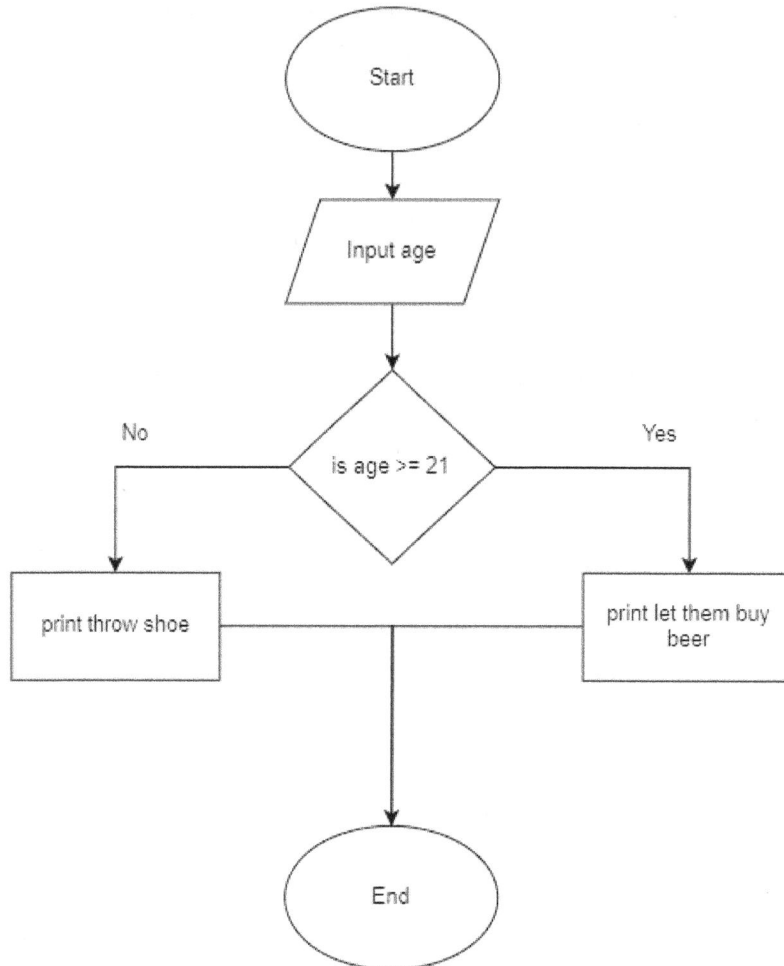

Figure 5.4 – Beer program flowchart

For this program, the flowchart is seemingly less complex than the pseudocode. In this case, we can represent the program with a single control statement. In the case of the flowchart, if the customer's age is less than 21, the program will display **throw shoe**, and if the customer's age is greater than 21, it will display **let them buy beer**. The one diamond symbol will represent the if and else if branches for the program.

Practice

Practice makes perfect. The best thing that a programming instructor can tell a student struggling with how to design an algorithm is to keep practicing. Unfortunately, that is easier said than done. How does someone practice designing algorithms? The answer is surprisingly simple. The best way to practice algorithm design is to take everyday tasks and convert them into program designs. Some examples of design tasks could be the following:

- Tying your shoes
- Cooking an egg
- Driving to the grocery store
- Starting your car
- Turning on a light
- Taking out your trash
- Packing lunch

The best thing a person can do to hammer in algorithm design is take simple, everyday tasks that they know well and perform daily and convert them into an algorithm. For this exercise, it is best to create a flowchart and corresponding pseudocode.

Performing these little design exercises will provide two benefits. The first is that it will provide practice designing algorithms. Second, it is also a common exercise for coding interviews. It is not unusual for a hiring manager to have a candidate pick an everyday task such as tying their shoes and have them design a pseudocode program, flowchart, or both that can represent the steps in the task. Therefore, getting into the habit of converting your everyday tasks into algorithms can very easily help land you a job.

By this point, you should have a decent grasp of designing algorithms. To round out the chapter, we're going to build one final project. We're going to design a program for a robot startup system.

Final project – Robot startup system

In automation programming, it is common to have to write software that can start a process, such as turning a robot on and getting it ready for operation. These systems can be quite complex and have many ins and outs. For this final project, we're going to design a system that can be converted to real PLC code for the startup of a stationary robotic arm.

Design requirements

For this system to work, the following steps must take place:

- If the robot is not powered on, it must be powered up.
- The robot has six joints that must be zeroed.
- The robot must go into wait mode.

This may seem like a simple program but don't be fooled, as very simple tasks can often have the most amount of gotchas.

For this project, assume the following:

- The system must perform a network check to ensure the robot can communicate its operations with other machines. This is a part of powering on.
- The robot needs a 5-minute warmup period.

The system needs to verify that proper voltage is being applied to the motors before it can operate. There is a light curtain that will stop the system in the event someone enters the operating area. These look like some surprise requirements. That's because they are. Very rarely will an engineer ever get something that is very straightforward. Typically, a task as simple as the one described in the initial bullets will not be that simple. Every project will usually have a series of necessary support tasks that will need to be carried out for the overarching process to properly work. So, as an engineer, be expecting these gotchas, and if it's your first time working on something new, be sure to ask if any additional support tasks need to be worked into the system.

Design logic

The first thing that we need to do is work out the necessary steps to carry out the operation. To do this, let's break the program down into a series of smaller operations and combine these in the overall design.

Power-up sequence

This is the sequence of steps for powering up:

- Power-up sequence

  - Send power-up signal to robot.
  - Pause the program for 5.5 minutes to ensure the 5-minute warmup period is completed.
  - Ping the network 40 times. If the network is unresponsive after 40 pings, throw an error message. If the network responds before 40 pings, continue with startup operations.

Zero robot joint

Since there are six joints in the robot arm, there are going to be six motors that we need to check. As such, we can use the following sequence:

- Zero sequence

  - Get voltage reading for motor 1.

  - Get voltage reading for motor 2.

  - Get voltage reading for motor 3.

  - Get voltage reading for motor 4.

  - Get voltage reading for motor 5.

  - Get voltage reading for motor 6.

  - Check if anything is in the light curtain.

  - Verify voltage readings are within limits. If everything is in parameters, home motors.

Wait sequence

The wait sequence will depend on the success of the previous two sequences:

- Wait sequence

  - Ensure the system is fully powered on

  - Ensure motors are zeroed

  - Wait for system commands

With all that established, we can now move on to writing the pseudocode for the program.

Pseudocode

With the logic laid out, let's work out the pseudocode:

```
//turn on phase
send on signal to robot.
pause 5.5 minutes
loop 40 times:
    ping network
    if ping received back
quit loop
if no ping response:
```

```
        throw error "Network not available"
//Zero phase
VoltsMotor1 = voltage on motor1
VoltsMotor2 = voltage on motor2
VoltsMotor3 = voltage on motor3
VoltsMotor4 = voltage on motor4
VoltsMotor5 = voltage on motor5
VoltsMotor6 = voltage on motor6
lightCurtain = light curtain detects movement
if VoltsMotor1 - VoltsMotor6 correct volts
    and lightCurtain detects no movement:
        rotate motors1 - motors6 until encoder reads 0
//Wait Sequence
Loop:
if ensure robot voltage > 0 and
ensure encoders position 0 and
command received:
    preform command
```

Compared to the other programs we've seen; this one is by far the most complex. Typically, an engineer would not use pseudocode for such a large and overarching algorithm. Instead, the engineer would use pseudocode to work out the smaller subsections of the program such as the zero sequence or the turn-on phase. However, as was seen in the preceding snippet, pseudocode can still be used to hammer out larger projects.

All things considered, a system such as the one we're designing is much better suited to be laid out with a flowchart. Due to the interworking sections that each have subsections of their own, a flowchart would be much more efficient in terms of laying out the design. Therefore, let's take a stab at designing with a flowchart.

Flowchart

For the flowchart, we could use something similar to *Figure 5.5*. The flowchart in *Figure 5.5* is almost one-to-one when compared to the pseudocode. Much like the other examples, there is a clear flow from symbol to symbol that marks the flow of the program. For the flowchart in *Figure 5.5*, the line that goes between the get light curtain command and the check is oriented upward. In the past examples, all programs flowed in a downward orientation; however, for this diagram, the arrow had to be oriented up to fit the diagram on a page. In terms of programmatic flow, nothing changed, and the program would still flow in a top-down, sequential fashion. The change in the orientation of the flow was merely a rendering decision and did not affect the program.

Now, in *Figure 5.5*, you may notice that the diamond symbol is used quite heavily. That is because the diamond can also represent a loop in flowcharting. Until this point, we have not used the symbol in that regard, so it may seem a bit awkward at first. However, if there is a line pointing back to the diamond, that command represents a loop. In this diagram, all loop diamonds have the word *loop* in them as well:

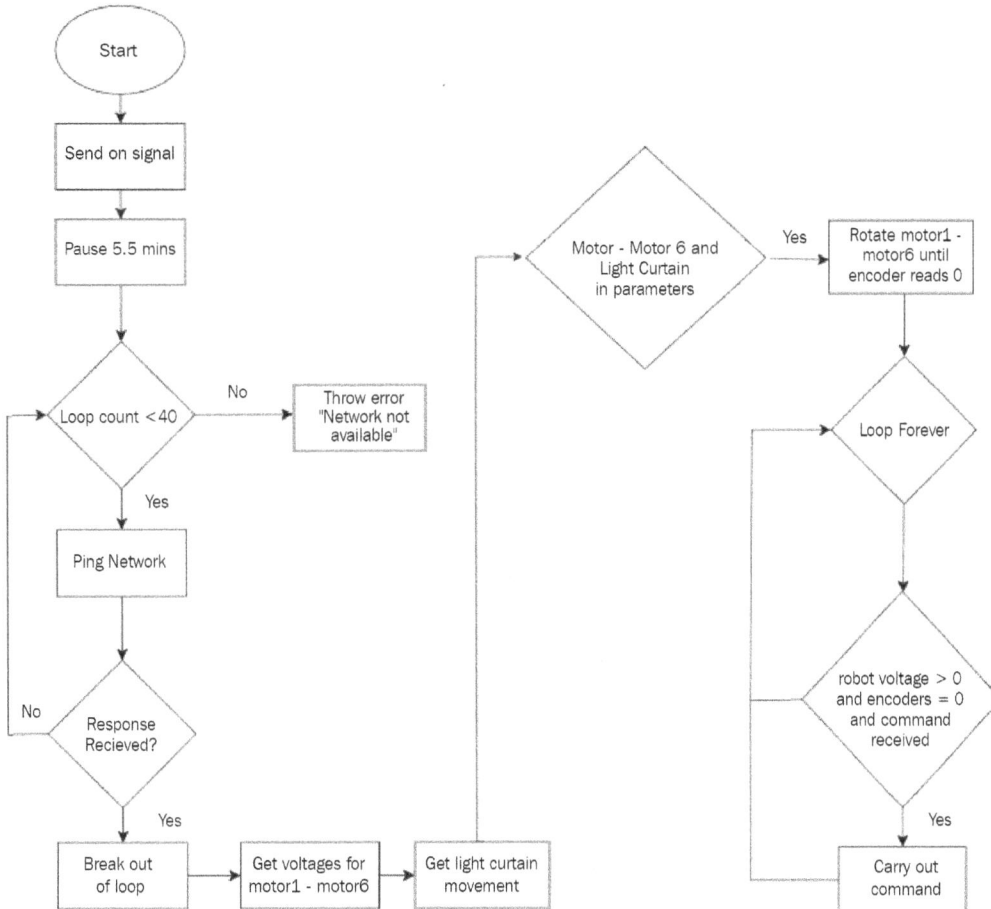

Figure 5.5 – Robot startup flowchart

As with the other examples, this is just one way of drawing the system. Much like with pseudocode, if the drawing makes sense and accurately depicts the system, an engineer can draw any way they wish. Also, note that this program does not have a true end. This was left out due to the program's forever loop. In this scenario, once the robot is turned on, it will loop forever; there will be no end to the program. The only way the program will restart is when the system is turned off, which was not factored into the design.

As stated before, for projects such as these that have many interworking parts, it is usually better to draft them with a flowchart. The flowchart seen here offers a more visual layout of how the system works, which means that it can be studied more thoroughly than just by reading copious amounts of pseudocode. The big reason for this is that this robot startup program is more than just a program; it is an overarching system that is composed of many smaller pieces that must work in unison. This means that what we're laying out is a high-level design. If this were a real-world project, the flowchart would be used to lay out the full system flow, and pseudocode would be used to draft out the individual sequences, as mentioned in the *Pseudocode* section. In all, the flowchart can be used later on to troubleshoot issues, find redundancies, and more, which can then affect the actual code designs.

Summary

This chapter explores ways to design a program and a high-level system. In this chapter, we have explored pseudocode and flowcharting, as well as techniques and programs that are used to render the designs. The important takeaway from this chapter is that a successful program does not start with code. Instead, a successful program starts with a decent design. A skilled engineer will spend a good amount of time with a pen and paper or with a rendering program, laying out the flow of a PLC program long before they touch a keyboard. This may seem oxymoronic and even frustrating to inexperienced PLC programmers. It is not uncommon for inexperienced engineers to forego the design phase of a project and jump into coding with a desire to build something. However, this attitude is the quickest way to get into programming trouble and have your fancy new machine doomed for the cyber scrap heap.

So far, we have explored a lot about program flow, hardware, and program design. In terms of becoming a quality engineer, we're well on our way, but before we can get into actual coding, we must make one more small detour in the next chapter and explore some basic logic functions to help us understand how our program works. In the next chapter, we're going to learn Boolean algebra.

Questions

1. What is a flowchart?
2. What is pseudocode?
3. What are the main differences between pseudocode and flowcharts?
4. List three programs that can be used to write pseudocode.
5. List three programs that can be used to draw flowcharts.
6. Why use a flowchart?
7. Can you draw flowcharts by hand?
8. What is a common interview technique that requires candidates to draw a design on a whiteboard?
9. What are the two functions of the diamond symbol in a flowchart?
10. Design a robot stop system that sequentially turns off all the systems we turned on in the final project.

6

Boolean Algebra

At first glance, this chapter may seem like the most intimidating because it revolves around math and logic. However, the material's bark is by far worse than its bite. The whole gist behind this chapter is learning to manipulate binary numbers and analyzing their outputs using what is called Boolean logic or Boolean algebra.

The term might be off-putting to most; however, the type of algebra that will be explored in this book is not the standard algebra that is taught in high school. Instead, the type of algebra that will be explored here is based on two numbers, 0 and 1. This type of algebra is extremely important to PLC programmers and even electrical engineers who work on digital systems such as PLC-based systems, as it allows engineers to model the logical behavior. More specifically, it allows engineers to more effectively model the behavior of switch configurations, contact configurations in Ladder Logic, and control statements in Structured Text.

In short, all modern electronics are, for the most part, digital in nature. Automation engineering is no different. Outside of the analog components that were explored in *Chapter 2*, almost all modern devices are either going to be digital or support a digital device. This means that understanding Boolean algebra is vital to the success of any automation programmer. Therefore, to begin our exploration into the world of Boolean algebra and logic, we're going to explore the following concepts:

- What is Boolean algebra
- Boolean operators
- Boolean expressions
- Solving Boolean equations
- Truth tables

To round out the chapter, we're going to apply what was covered in it and create a truth table from scratch.

Technical requirements

This chapter will not require any special software and has no other technical requirements. It will be somewhat math intensive, but no prior experience with Boolean algebra is required.

What is Boolean algebra?

Boolean algebra sounds like a scary form of mathematics, but it's actually quite simple by nature. As stated in the introduction, Boolean algebra has little to do with the algebra that is normally taught in high school or college. The goal behind Boolean algebra is to perform a series of operations on a 0 or 1 to determine whether the output of an equation is going to be either 0 or 1.

If one really wanted to split hairs, there is a difference between Boolean algebra and Boolean logic. If one were to Google the difference, one would find that a lot of the definitions overlap. Even though there's a slight difference between Boolean algebra and Boolean logic, in everyday speech, the terms are used interchangeably. In this book, we're going to follow the conversational norm and use the terms interchangeably.

Boolean algebra is often used to model the behavior of electrical and software systems. This is because most sophisticated hardware systems are, at their heart, simply a series of on and off switches baked onto a chip that is controlled via a program. Due to this nature, engineers will represent "on" with the digit 1 and "off" with the digit 0. Engineers can then use Boolean algebra to model their circuit or program's behavior based on a given series of inputs.

All things considered, a PLC programmer, or a programmer in general, does not spend hours of their time crunching Boolean equations. Instead, quality programmers will use Boolean equations to help work out things such as truth tables and model complex control statements to understand, design, and test complex switch configurations and control statements to save time in the long run. Therefore, having a basic understanding of Boolean algebra and how it relates to programming is important for any developer in any field. To be a quality programmer, a developer will need to know how to read a truth table to understand basic Boolean operators, be able to evaluate "complex" Boolean equations in control statements, and so on. Understanding Boolean algebra is even more prevalent for automation engineers due to the electrical element that is mixed in with PLC-based software.

As you're reading this chapter, don't be concerned if this doesn't come naturally. Do not simply gloss over this chapter either, because there will come a time in your career as an automation engineer when you need to use these techniques to either design, test, or troubleshoot a system. With all that being said, let's explore the basics of Boolean algebra: logical operators.

Boolean operators

In real-world settings, very few things work in a one-to-one fashion, for example, flipping on a single switch to power on a machine. Even a thing as simple as turning on a light often requires a person to flip two switches! In other words, to turn things on or off, multiple conditions may need to be satisfied.

For example, an output as simple as a lightbulb may need at least two switches to be on to power the bulb. Essentially, all this can be boiled down to three basic Boolean operators, which are as follows:

- AND
- OR
- NOT

The basic operators

A Boolean operator is a word that signals an operation on one or more conditions. For many, this explanation may not make a lot of sense, and the best way to demonstrate it is with a few examples. An easy way to understand operators is to examine a couple of sentences.

The AND operator

The AND operator signals that all conditions must be true for the final output to be true. For example, consider the following sentence:

Both switch 1 AND switch 2 must be on for the light to be on.

If one analyzes this sentence, there are two inputs, which are switch 1 and switch 2, and the output is the state of the light bulb (on or off). In this case, the AND word signals that both switches have to be on for the light to be on. If this were converted to pseudocode, it would look like the following:

```
If switch1 = on AND switch2 = on:
    Turn on lightbulb
```

Overall, the AND operator is very common in the automation world. Many times, certain processes must be on for another process to start. In these cases, engineers will use the AND operator to accomplish this.

The OR operator

The OR operator is kind of like the AND operator; however, it works a little differently. Where the AND operator requires all the inputs to be true, the OR operator only requires one or more of the inputs to be true for the output to result in a true state. With that, consider the following example:

Either switch 1 must be on OR switch 2 must be on for the lightbulb to turn on.

In this sentence, for the lightbulb to be on, either switch 1, switch 2, or both switches must be on for the lightbulb to be on. In pseudocode, this would look like the following:

```
If switch1 = on OR switch2 = on:
    Turn on lightbulb
```

Much like with the AND operator, the OR operator is equally common. In automation, it is also very common to require one or more processes to be on (or true) before another process can be started. In these cases, engineers will use the OR statement to accomplish that.

The NOT operator

Compared to the AND or OR operator the NOT operator works a little differently. Unlike AND or OR, which take multiple inputs, NOT only takes one. Also, unlike the other two operators, which require at least one input to be on for the output to be on, the NOT operator requires the input to be OFF for the output to be on and vice versa. In other words, the NOT operator inverts the input. In terms of sentence structure, a NOT operator would be something like the following:

To turn the light on, the switch must NOT be in the on position.

In this case, for the light to be turned on, the switch must be off. In terms of pseudocode, it would look something like the following:

```
If switch1 = Off
     Turn light on
If switch1 = on
     Turn light off
```

A common use case for the NOT operator is with an emergency stop switch, or, as they are more commonly known as **E-Stops**. Typically, if an E-Stop is engaged or the switch is "on," the machine will usually stop. On the other hand, if the switch is in its naturally disengaged state, the machine will turn on. This is the inverse at work; essentially, for the machine to be on, the switch must be off, and for the machine to be off, the switch must be on.

Operators are very powerful in programming. Thus far, we have only examined operators in their purest form: AND, NOT, and OR operators. Though these are often used alone, these operators can be used in conjunction with each other to form more complex logic. Moreover, the outputs for these complex sequences can be hard to figure out without some basic math. As such, the next section is going to explore the concept of Boolean expressions and how multiple logic operators can be used in tandem with each other.

Boolean expressions

Often, whether a control statement evaluates to true and a machine properly turns on will be dictated by a number of conditions, such as a series of variables being true or a series of switches being on. Boolean expressions are essentially specialized math equations that will determine whether an output will be on or off based on the state of the inputs.

The whole "math" concept may frighten a lot of people; however, the math behind Boolean expressions is not scary at all and is little more than adding or multiplying 1s and 0s. In a mathematical sense, each of the Boolean operators we explored previously will boil down to addition, multiplication, or simply flipping a value from 0 to 1, or vice versa. This, for the most part, means, that if you can multiply by 1 or 0 and add by 1 or 0, you have a solid foundation for Boolean calculations.

Exploring NOT

To start off with Boolean expressions, let's explore the NOT operator in more depth. As was explored earlier, the NOT operator simply means inversion. The NOT operator is commonly denoted with the ¬ symbol. When you come across this symbol, all you must do is flip the value. Consider the following example:

A = 1

Then,

¬A = 0

As can be seen, in the example, if the variable A = 1 then ¬A is 0. Now consider the next example:

B = 0

Then,

¬B = 1.

Again, all we had to do was toggle the value.

Double negation rule

Now, much like traditional algebra, Boolean algebra has a set of laws that one can use to help solve equations. In terms of the NOT or negation operation, a very important law is the double negation rule. Consider the following:

If A = B, then ¬ ¬ A = B

Essentially, a double negation has no effect on the variable that the NOT operator is being performed on. With that, consider the following:

If A = 1 then ¬ ¬A = 1

If you think about it, this is valid because if we perform the first negation on 1, it becomes a 0. Then, if we perform a negation on 0, it reverts to 1. So, in the end, the value of A never changed.

Expanded double negation rule

If you reflect on this rule some more, you'll find that if you have an odd number of negations, the final value will be inverted, but if you have an even number of negations, the final value will be equal to the original value. Consider the following:

A = 1

¬ ¬ ¬ ¬A = 1

Because

¬1 = 0 -> ¬0 = 1 -> ¬1=0 -> ¬0 = 1

Now, if we negate a variable three times, the value will be inverted. To demonstrate this, consider the following:

A = 1

¬ ¬ ¬A = 0 because,

-1 = 0 -> ¬0 = 1 -> ¬1 = 0

Therefore, the output for the triple negation was 0.

The OR operator

In terms of Boolean mathematics, the OR operator can be tied to addition. OR can be any of the following:

- 1 + 1 = 1
- 0 + 0 = 0
- 1 + 0 = 1
- 0 + 1 = 1

In the case of the bullets, the numbers on the left of the equal sign represent inputs, while the numbers to the right of the equal sign represent outputs. These inputs can represent switches in a circuit or an input from a PLC module. This concept will be true for any of the operators, such as OR, AND, or NOT.

When it comes to the OR operator, a shortcut to figure out whether the output should be 1 or 0 is to simply examine the input, or in this case, the numbers being added. If any of the numbers in the equation is 1, then the output is also 1. *This is a handy trick to understand when it comes to PLC programming, especially when you start working with control statements.*

As you're reading along, it is important to memorize the bullets, as these bullets are a basic truth table. If you're not familiar with truth tables, they will be covered in more depth in the *Getting to know truth tables* section of this chapter. For now, let's explore the AND operator

The AND operator

When one sees the AND operator, it usually signals a multiplication operation. Here's an example:

- $1 * 1 = 1$

- $0 * 1 = 0$

- $1 * 0 = 0$

- $0 * 0 = 0$

For the AND operator to have an output of 1, all inputs must be 1. Thinking back to basic multiplication, anything that is multiplied by 0 is always 0. In terms of the AND operator, this simple mathematical principle is very handy in figuring out the output. Another shortcut trick that comes in handy with the AND operator is to look at all the inputs. If one or more inputs are 0, then the output will automatically be 0.

Much like in mathematics, there are laws and identities that govern how these operations behave. As such, the next section is going to explore some of the more basic identities and laws.

Operator laws

There are a number of laws that govern logic operations; however, we're only going to cover the following: identity law, idempotent law, and commutative law. There are many more laws that can be used. For now, though, we're only going to explore the more basic laws to get familiar with things.

Identity law

Similar to how a person has an identity, so do mathematical operations. The identity law is a Boolean law that dictates that when a term is ORed with 0 or ANDed with 1, the result will always be equal to the term. For the OR operator, the following identity can be assumed:

$A + 0 = A$

For the OR operator, any value with a false input will always be the other input (A). For example, if A = 0, then the output will be 0. If A is 1, then A will be 1.

The identity rule for the AND operator can be summarized as follows:

$A * 1 = A$

In other words, if one of the inputs is 1, then the output will always be A. As such, if A = 0, then the output will be 0; if the input is 1, then the output will be 1.

Idempotent law

The idempotent law is a fancy way of saying that if all the inputs of a logical equation are the same, then the output reflects the inputs. In other words, the idempotent law can be boiled down to the following for the AND and OR operators:

- OR Operator: A + A = A

- AND Operator: A * A = A

Essentially, this means that any term ANDed or ORed with itself is always equal to itself. All things considered, this is a very easy-to-grasp rule. With that, we're going to explore one final law, the commutative law.

Commutative law

The commutative law deals with the order of variables. By that, what is meant is the way the variables are arranged in the equation. Keeping with the two-input theme, the commutative law will boil down to the following:

- A + B = B + A

- A * B = B * A

With all that we have learned thus far, let's look at a few examples, crunch some numbers, and work on solving Boolean equations.

Solving Boolean equations

Practice makes perfect. Just as many people learned to master traditional algebra by solving practice problems, we're going to look at a couple of examples to practice solving Boolean equations.

Examples

The following are a couple of examples of how to solve Boolean equations. For most of these practice problems, there are going to be three inputs.

Example 1

Calculate the output for the following equation:

$$(A + B) * C$$

Assume A = 1, B = 0, and C = 0.

In terms of PLC programming, this equation would correspond to a system with three inputs. It could represent a switch circuit, or it could represent a series of Boolean values in a control statement. Either way, the math is going to be the same.

To solve this, let's first break it down. Since A + B are enclosed in parentheses, these can be thought of as individual inputs for a special configuration. Since there is a plus sign, these two variables can be said to be in an OR configuration. So, what we can do is abstract that out and re-write the equation as follows:

$$D = A + B$$

Now, we can replace A + B with D in the original equation, which will yield the following:

$$D * C$$

From here, we have the variables in an AND configuration, and as stated above, C is 0. So, we can plug in 0 for C and we'll get the following:

$$D * 0 = 0$$

Since C was 0, and the variables that were in an OR configuration were in turn in an AND configuration with C, the output of the equation will be 0.

Example 2

For our next example, let's solve the following equation:

$$(A + \neg B) * (C + B)$$

For this equation, let's assume A = 1, B = 0, and C = 1. To solve this equation, let's break it down into two smaller equations and combine them at the end. So, the first thing we're going to do is the following:

$$D = (A + \neg B)$$

Then, the following:

$$E = (C + B)$$

The next thing we're going to do is solve for D. As such, plugging in the numbers will yield the following:

$$D = (1 + \neg 0)$$

$$D = 1 + 1$$

Therefore,

$$D = 1$$

Now, we can solve for E:

$$E = 1 + 0$$

Therefore,

$$E = 1$$

Now, we need to solve the original equation. Remember that the original equation is now D * E or D AND E, which will yield the following:

$$1 * 1 = 1$$

Therefore, the output for the equation will be 1.

By now, we should have a decent grasp of how to solve basic Boolean equations. Once we solve a Boolean equation, we need a way to conveniently convey what the outputs should be based on the inputs. A simple way to do this is with what's called a truth table.

Getting to know truth tables

What are truth tables? In short, they are easy-to-read tables that show all the outputs for all the combinations of inputs. In this section, we're going to explore the basic truth tables for the standard operators as well as for custom Boolean equations.

Basic operators

The following are the truth tables for the standard operators. Typically, you shouldn't have to worry about memorizing complex truth tables. However, of all the truth tables, these are the ones you should memorize because they correspond to basic operators.

NOT table

The first truth table that we need to memorize is the table for the NOT operator:

Input	Output
1	0
0	1

Table 6.1 – NOT truth table

Reading truth tables is quite easy. All one needs to do is read each row. The rows will show the state of the input(s) and the corresponding output. For example, if one skims the first row, they can easily see that when the input is 1 or True, the output will be 0 or False. If one were to read the second row, one would see when the input is 0 or False, the output will be 1 or True.

Once you have memorized the NOT table, you may proceed to the AND operator.

AND truth table

The next table that should be memorized is the AND truth table, which can be viewed in *Table 6.2*:

Input 1	Input 2	Output
1	1	1
0	1	0
0	1	0
0	0	0

Table 6.2 – AND truth table

Notice that the AND table has two columns whereas the NOT table only had one. This is because, as we have seen throughout the chapter, the NOT operator only depends on one Boolean value, whereas AND and OR will depend on at least two. This means that each row on the truth table will need to account for the extra input.

OR truth table

The final basic truth table we're going to explore is for the OR operator. This truth table can be viewed in *Table 6.3*:

Input 1	Input 2	Output
1	1	1
0	1	1
1	0	1
0	0	0

Table 6.3 – OR truth table

The AND and OR operators are not constrained to only two inputs. These operators can have any number of inputs. Typically, these operators have two or more inputs, or Boolean values, that the output depends on. In terms of digital logic chips, an engineer typically will not see a chip with gates that are more than three inputs.

The tables presented so far should be memorized, as these tables are core to writing effective PLC code. However, in common PLC programming, control statements are sometimes more complex than just simple AND and OR operations. As such, an important skill that we need to develop is learning how to construct truth tables for custom Boolean equations. In the next section, we're going to combine what we've learned and create a custom truth table from scratch!

Final project: Creating custom truth tables from scratch

For the final project, we're going to create a truth table for a Boolean equation. Before you can build custom truth tables, you need to thoroughly understand how to solve Boolean equations and understand how the truth tables work from the last section. For this section, we're going to create a truth table for the following equation:

$$(A*B) + B$$

To begin building this truth table, we first need to figure out how many rows are needed to complete the table. To figure this out, we need to count the number of inputs or variables, as the number of rows is equal to the following equation:

$$2^n$$

Where n is the number variables or inputs for the equation. Since the equation only has variables A and B, we need the following number of rows:

$$2^2 = 4$$

So, there will be four rows that we need to calculate. A good strategy for creating the truth table is to insert all the input combinations first, as this will make keeping track of things easier. As such, the first thing we're going to do is lay out the input columns for the truth table:

A	B	Output
1	1	
1	0	
0	1	
0	0	

Table 6.4 – Truth table with no output

Now that we have a skeleton of a truth table, we can start to plug in the values for A and B.

Row 1

For row one, we're going to plug in 1 for both A and B. With that, we can rewrite the equation to match the following:

$$(1*1) + 1$$

Let's break down the equation into pieces again:

$$(1 * 1) = 1$$

Now, we can simplify the equation again to the following:

$$1 + 1$$

Since this is an OR statement, the final output is 1 because both inputs are true. So, the output for the first row is 1.

Row 2

The next row has A set to 1 and B set to 0. This means that the equations can be rewritten to match the following:

$$(1*0) + 0$$

Simplifying this, we get the following:

$$(1*0) = 0$$

$$0 + 0 = 0$$

As such, the output for the second row is 0 or false.

Row 3

For the third row, we're going to invert the values for A and B so that A will be 0 and B will be 1. Plugging in these values will render the following:

$$(0*1) + 1$$

Simplifying this down, we get this:

$$0 + 1 = 1$$

Which means that the third row will be true.

Row 4

The last row will be the easiest to calculate. In this case, both A and B will be 0. As such, if we plug in 0 for both variables in the equation, we get the following:

$$(0*0) + 0$$

In this case, since there are no negations and there are no true or 1 values, the result will be 0. Therefore, we can say that the output for the last row will be 0.

Final truth table

Now that we have all the outputs for the table computed, we can finish the truth table:

A	B	Output
1	1	1
1	0	0
0	1	1
0	0	0

Table 6.5 – Truth table with outputs

As can be seen, the only time the output for the equation will be true is when both outputs are true or when B is true and A is false. Also note that the result of the equation will mirror the value of B.

Summary

In this chapter, we explored Boolean algebra, or, as it is sometimes called, Boolean logic. We have explored the basic operators, how to compute Boolean values, truth tables, and more. In short, this chapter may have seemed unnecessary; however, knowing how to calculate Boolean equations and create truth tables is a prerequisite to becoming a quality PLC programmer.

This chapter is the last in the design/logic section. Starting in the next chapter, we're going to move on to more practical aspects of PLC programming and start writing working PLC code. As such, the next chapter will focus on getting CODESYS up and running and writing our first PLC program!

Questions

1. Solve the following equation: $(A + B) + A$ where $A = 1$ and $B = 0$.

2. Write the truth table for question 1.

3. What is the truth table for the AND operator?

4. What does 0 represent?

5. What does 1 represent?

Further reading

- Boolean Algebra:

  https://www.geeksforgeeks.org/boolean-algebra/

- Boolean Algebra:

  https://www.tutorialspoint.com/computer_logical_organization/boolean_algebra.htm

Part 2: Introduction to Structured Text Programming

This part is an introduction to Structured Text PLC programming. The goal of this part is to set up the programming environment and become familiar with Structured Text programming. This part will be very technical and in-depth with plenty of challenges and critical thinking problems to help apply what you learn. The core tenets of this part are understanding Structured Text, implementing `IF` statements, using the different types of loops, learning how to use built-in function blocks, performing calculations, and more.

This part has the following chapters:

- *Chapter 7, Unlocking the Power of ST*
- *Chapter 8, Exploring Variables and Tags*
- *Chapter 9, Performing Calculations in Structured Text*
- *Chapter 10, Unleashing Built-In Function Blocks*
- *Chapter 11, Unlocking the Power of Flow Control*
- *Chapter 12, Unlocking Advanced Control Statements*
- *Chapter 13, Implementing Tight Loops*

7
Unlocking the Power of ST

Alright, now it's time for the fun stuff! Now that we have a solid foundation in program design, computer science, hardware, and so on, we can get into the exciting material and start writing code. This chapter is going to be dedicated to installing the software that you'll need to write PLC code, as well as taking an in-depth look at **Structured Text (ST)**.

ST, is, for lack of a better word, feared by many PLC programmers, as it can be intimidating. However, ST does not have to be something to be afraid of. In fact, for many applications, ST can be easier to use and understand than the more popular **Ladder Logic (LL** or **LD)** programming language. What you'll find in this chapter is that ST bears a very close resemblance to the pseudocode we used in *Chapter 5*, so if you skipped over that chapter or didn't fully grasp the information presented, it's a good idea to return to it until you feel comfortable with that material.

So far in the book, we've explored the basics of some of the PLC programming languages such as LL, and we've done a high-level comparison with ST. However, we have yet to fully explore the benefits of ST and why we should use it over the other languages that the IEC 61131-3 standard governs. That is going to change in this chapter, as we're going to explore the following concepts to really hammer home when and where ST should be used.

In this chapter, we're going to explore the following topics:

- What is ST
- Why ST is important
- ST versus LL
- What CODESYS is
- How to install CODESYS

Finally, to round out the chapter, we're going to create the first program that every programmer writes in a new language, the famous Hello World program.

This chapter cannot be skipped, and all examples in this chapter must work before you can move on to the subsequent chapters. Therefore, it is worth spending a little extra time and effort on this chapter to ensure you have the material down; otherwise, the rest of the book won't make much sense.

Technical requirements

This chapter will require a working copy of CODESYS to be installed on your machine. You can download CODESYS at the following link:

https://store.codesys.com/de/

CODESYS is free to download and use. It has a built-in simulator that can run your code without the need for a PLC. As such, it is not only a great PLC programming tool but also an awesome learning tool!

For your reference, all code examples can be retrieved from the following GitHub URL:

https://github.com/PacktPublishing/PLCs-for-Beginners

What is ST?

To begin our exploration into the world of ST, we first need to know what ST is. Put bluntly, ST is a text-based programming language that allows automation programmers to write PLC code in an easy-to-read-and-follow format that resembles natural language. In other words, it allows PLC programmers to write code using actual words as opposed to symbols.

ST most closely resembles the old BASIC programming language, with a touch of Ada thrown in for good measure. Overall, ST gives developers a very natural and friendly flow to reading and writing software. To utilize ST effectively, all you need to do is memorize certain keywords, as opposed to memorizing exotic symbols, as you would normally do in Ladder Logic programming. So, what does ST look like?

Area of a circle program in ST

For the inexperienced, ST may seem like an exotic concept that is fraught with perilous complexities. However, ST is quite simple. The following ST program will calculate the area of a circle with a radius of 4. As can be seen, there are two blocks that each file will have – a variable block and what can be thought of as a logic block. The variable block for this program will look like the following:

```
PROGRAM PLC_PRG
VAR
     radius : INT := 4;
     area : REAL;
END_VAR
```

This code block has two variables that were created by us, the programmer. The first variable is called radius and has a data type of INT. If you don't know what that means that's fine, we're going to cover that in the next chapter. The second variable is called area, and it has a datatype of REAL. Again, if you don't know what that means, don't worry – we'll cover it in the next chapter. Both variables were created and named by us. The variable names are just symbols. We could have named them literally anything, such as Bob or Larry. However, in programming you want your variable names to be reflective of what their job is in the program, so we opted for the names radius to house the size of the circle's radius and area to house the computed area, as they logically reflect what they do in the program.

As for the "logic" block, it should look like the following:

```
area := 3.14 * expt(radius, 2);
```

This block will carry out the program's operations. In other words, this is where the program does what it's supposed to do. In this case, it takes the radius value from the variable block, squares it, and then multiplies it by pi. As a result, when the program is run, it produces the output in the following screenshot:

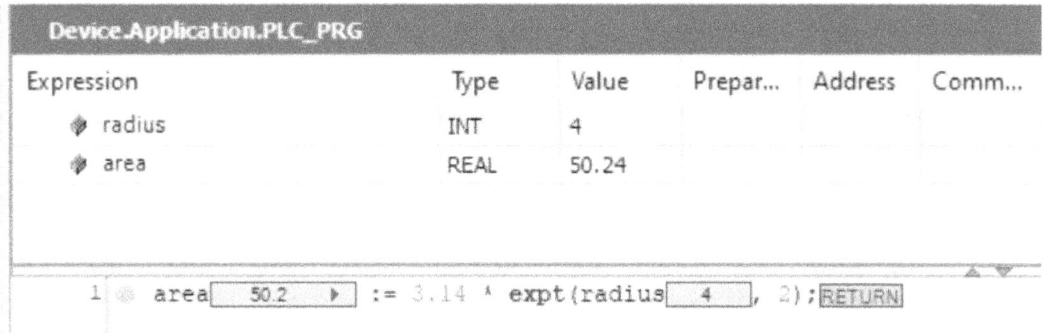

Device.Application.PLC_PRG					
Expression	Type	Value	Prepar...	Address	Comm...
radius	INT	4			
area	REAL	50.24			

```
1   area   50.2   ►   := 3.14 * expt(radius   4   , 2); RETURN
```

Figure 7.1 – The circle area program output

Now, this example, much less the inner workings of the code, probably doesn't make any sense, and for now, that's fine. The main takeaway is to demonstrate how close ST is to the English language as well as the pseudocode we explored in *Chapter 5*.

Now, 90% of all PLC projects will use LL. For many PLC programmers, LL will be enough, and ST would be largely unnecessary. So, a logical question that a PLC programmer may ask themselves is, why should we care about ST, and overall, why does ST matter? The next section is going to answer those questions!

Why is ST important?

If you've ever worked as a PLC programmer, you're probably wondering, why bother with ST? Why not just stick with time-tested LL? The truth is that LL is an excellent programming interface for relatively straightforward projects. However, for complex projects, such as those that require advanced math calculations or other applications that have high complexity, ST shines.

To put it bluntly, the way ST can simplify a program still doesn't answer why it is important. In truth, the future of PLC programming is ST. By no means will LL go extinct, at least not anytime soon. However, the future of automation will incorporate technologies such as artificial intelligence, IoT, and other advanced concepts that will be very awkward to incorporate into LL-based programs. In other words, ST is important because it will be more capable of handling future-based technologies. As time ticks on, new technologies such as IoT and artificial intelligence will become more prevalent, which are technologies that LL was never designed for.

The following summarizes why ST is important:

- **Reduced complexity**: ST can drastically reduce the complexity of a program. For applications that incorporate new technologies, such as artificial intelligence, cloud technologies, and IoT, it is often much easier to integrate those technologies with ST than it is with LL.

- **Ease of troubleshooting**: This relates to reduced complexity, but it is worth mentioning. ST, due to its natural language flow, makes it much easier to read code, especially when the program incorporates complex math such as calculations for motion control.

- **No need to understand complex symbols**: ST is built on natural language. This means that there is no need to understand what rungs are, how to interact with contacts and coils, and so on. Anyone who can read can usually follow along with an ST program. Although many inexperienced automation programmers view ST as an exotic and intimidating programming interface, in reality, it is often much easier to understand using the material that was presented in the previous chapters.

Now, these are just a few reasons why ST is important and why it will be more important in the future. However, its importance cannot be summarized fully in three bullet points. For now, we're going to move on and really explore ST by comparing it to LL!

ST versus LL

As has been implied before, ST is often viewed as either unnecessary, complex, downright complicated, or unimportant to many inexperienced or non-formally trained programmers. This is a very misleading perception that has been encouraged by many years of programmers becoming a little too comfortable with old-school LL programming. In all, a lot of the fear of ST stems from many companies and engineers being unwilling to step out of their comfort zones and try ST on a new project. This section is going to try to dispel that myth with examples.

To really understand ST and how it relates to LL in terms of complexity and usage, we need to compare it to LL. To do this, we're going to look at a few examples. The first example is the area of a circle program we wrote in ST previously to see how the code bases compare.

Example 1 – The area of a circle program – LL

To keep things organized for the LL version of the program, you will need the following variables:

```
PROGRAM PLC_PRG
VAR
     radius : INT := 4;
     pi : REAL := 3.14;
     raduis_squared : REAL;
     area : REAL;
     enable : BOOL := TRUE;
END_VAR
```

As you can see, this version of the program requires more variables. At a minimum, the enabled variable must be added to enable the multiplication blocks. Since there isn't a standard exponent block, we also need an extra variable (`radius_squared`) to accommodate that. The last extra variable that we created for this example is the `pi` variable. This variable was not included in the ST version of the program and is not technically needed here either. However, it was added to make the value assignments in the multiplication blocks a bit easier to follow.

> **Note**
>
> This example uses pure LL. Depending on the programming environment, there are some LL blocks that allow users to run custom ST code. However, the key here is to understand that ST is still needed!

Now that we've explored the variables for the program, we need to look at the logic for the program. As you can see in *Figure 7.2*, the logic is a bit more complex than what can be seen in the ST version:

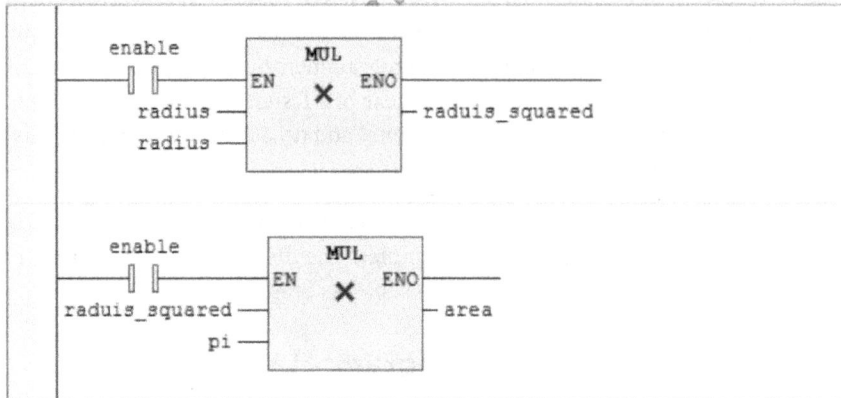

Figure 7.2 – The area of a circle LL version

In this version, there is an extra contact that needs to be enabled to activate the two blocks. On top of that, the calculation is broken up across two rungs, and the variables assignment on the multiplication blocks is a bit more awkward to assign than in the ST version of the program.

When the program is run, it will result in *Figure 7.3*:

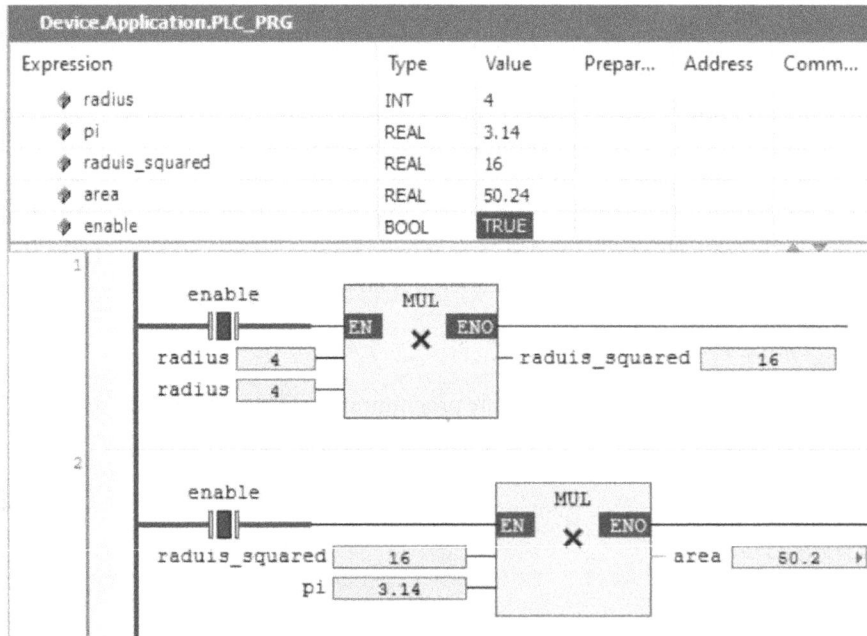

Figure 7.3 – The area of circle LL execution

When this program is executed, the area value is 50.2, just like with the ST version of the program. In terms of execution and results, there were no differences. All that changed was the overall complexity of the program itself. The LL version is less intuitive to follow, requires more code, and is simply more awkward.

> **Note**
>
> This code is for demonstration purposes only. It is not meant to be run, but the code can be downloaded from GitHub to be explored. If you do not understand the inner workings of the program or any subsequent programs in the chapter, don't fret – that material will be covered throughout the various chapters.

Now, one simple example does not do justice to the difference between the two interfaces. To fully appreciate the difference, a couple more examples need to be presented. Let's compare and contrast an LL program and an ST program that are responsible for turning on and off a light.

Example 2 – Toggling a light

In this next example, let's explore a program that can turn on and off a light. The program will be akin to a digital switch. When the proverbial switch is turned on, so too will the light; when the switch is turned off, so too will the light.

The LL code

The first program we're going to write is the LL version. As such, the first piece of code that we're going to need to produce is the variables section:

```
PROGRAM PLC_PRG
VAR
    switchState : BOOL := FALSE;
    lightState  : BOOL;
END_VAR
```

For this program, there will only be two variables, called swtichState and lightState. The switchState variable will serve as the "digital light switch." The lightState variable will mirror the state of the switchState variable.

In terms of the logic for this program, see *Figure 7.4*:

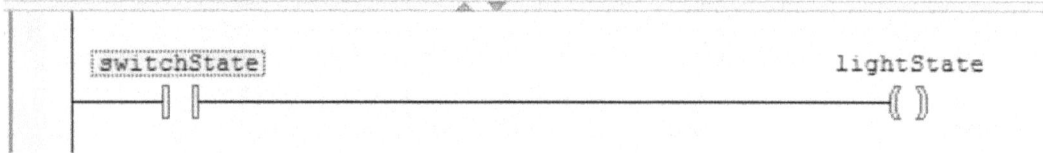

Figure 7.4 – The LL light switch

This program is composed of a contact and a coil. A contact is like a switch, and a coil is an output like a light. For this program, when the light switch is **on**, so is the light.

When the program is executed, the `switchState` variable will be set to `False`, which will result in *Figure 7.5*:

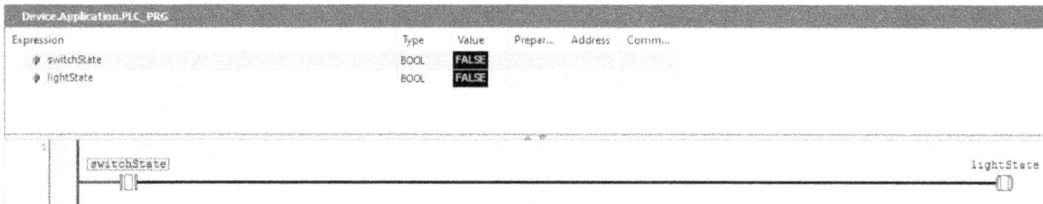

Figure 7.5 – LL – The light is off

If the light is switched on, the result will be *Figure 7.6*:

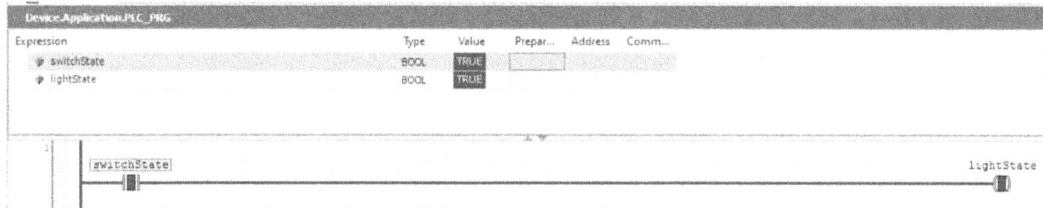

Figure 7.6 – LL – The light is on

If you're reading this book, there is a decent chance that you may not know how to program in LL. On the other hand, if you are familiar with LL, think back to when you were first learning it. If your boss dropped this program in your lap, would you have any idea what is going on in this program? Would you be able to follow it without Googling what LL is? Would you even be able to tell whether the coil is energized? Chances are, probably not at first glance. With that, let's explore the ST version of this program.

The ST code

Now that we've explored the LL version of the program, let's explore the ST version. To begin the program, we're going to recycle the same variables. As such, for our ST program, the variable section should look like the following:

```
PROGRAM PLC_PRG
VAR
     switchState : BOOL := FALSE;
     lightState  : BOOL;
END_VAR
```

For this version of the program, the variables will serve the exact same purposes. Therefore, we can move on to implementing the logic, which is as simple as the following code block:

```
lightState := switchState;
```

This simple statement is all that is needed to convert the LL program to ST. When this program is run and the switch variable is set to `True`, it will produce the output shown in *Figure 7.7*.

As can be seen, the `lightState` variable directly mirrors the `switchState` variable. The only real difference between the two versions of the program is that there isn't a blue line that denotes whether the output or, in this case, the `lightState` variable is energized or not. In the case of *Figure 7.7*, the only way to tell whether the light is on is by reading the `Value` column or the state next to the variable:

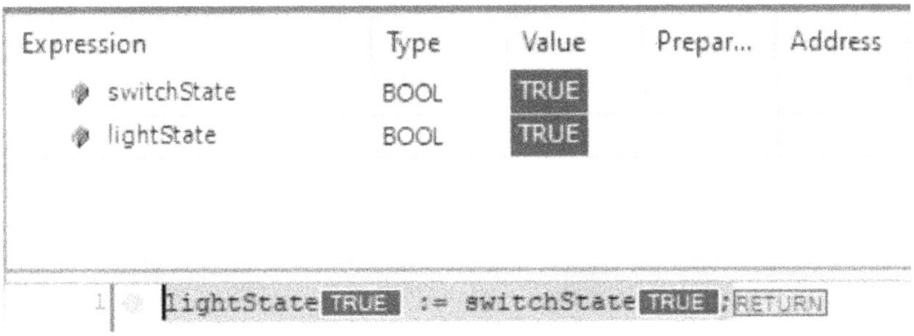

Expression	Type	Value	Prepar...	Address
switchState	BOOL	TRUE		
lightState	BOOL	TRUE		

```
1   lightState TRUE := switchState TRUE ;RETURN
```

Figure 7.7 – ST – light on

When the light switch is turned off, the program output should look like *Figure 7.8*:

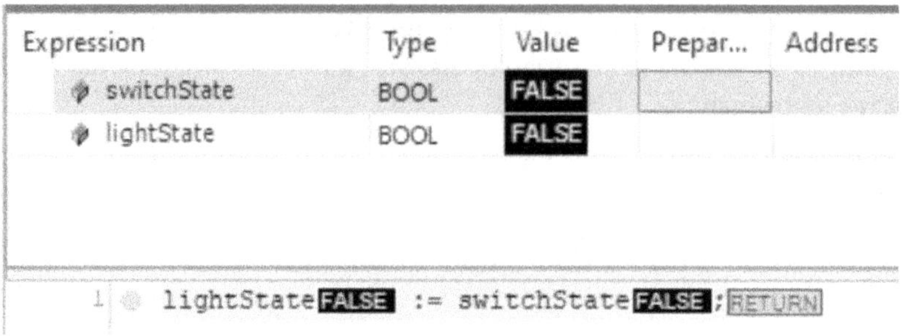

Expression	Type	Value	Prepar...	Address
◈ switchState	BOOL	FALSE		
◈ lightState	BOOL	FALSE		

```
1   ● lightStateFALSE := switchStateFALSE;RETURN
```

Figure 7.8 – ST – light off

The outputs from both programs are essentially the same. Assume you know nothing about programming or PLCs. Which program do you think will be easier to follow or write? The LL version, which has a bunch of non-intuitive commands that a layperson would not readily understand, or the ST version, which is basically an equal sign? Chances are you would probably lean toward the simplicity of ST.

There are a lot more examples that can demonstrate the differences. However, let's summarize the differences and similarities:

- **Syntax**: Much like any other written language, ST has certain rules and grammar to follow for the program to compiler and run. Conversely, LL has rules as well, but the biggest difference is that ST will use written text whereas LL will use symbols. Both languages have their own learning curves. Depending on your background, LL may be easier to learn at first; however, ST is more future-proof.

- **Nature**: LL resembles electrical circuits because it is derived from old-fashioned relay logic. Conversely, ST is a word-based programming language. In LL, the programmer will need to memorize and understand what certain symbols do, whereas in ST, the programmer must memorize and understand what keywords do. Again, if someone has a background in electronics or as an electrician, LL may be easier to learn upfront, but they will still need to learn ST to future-proof their programs.

- **Ease of use**: As we saw in the example, and as opposed to traditional myth, LL typically requires more rungs and code configuration to accomplish the same results as an ST program. The ST programs were typically shorter and easier to follow.

Again, these bullets are like the bullets listed previously. Overall, these are just some high-level attributes that contrast the two programming interfaces. Both lists are by no means comprehensive.

A logical question at this point would be, how can an engineer write a PLC program? In the following section, we're going to explore the necessary software that will be needed to write ST and get some hands-on practice. To do this, we're going to explore a programming software called CODESYS.

What is CODESYS?

If you went to or are currently going to school to be an automation engineer, you're probably familiar with something such as RSLOGIX or S7 PLCs and programming suites. It can be forgiven if entry-level professionals and students were to believe that these were the only programming environments on the market. However, they would be gravely mistaken.

Enter the world of CODESYS. CODESYS is an extremely powerful programming environment that more closely resembles a standard IDE such as Visual Studio or PyCharm than a PLC programming suite. CODESYS is a free-to-download-and-use development package that has a built-in simulator, allowing developers to run their code without the need for actual PLC hardware. This makes CODESYS one of the premier PLC programming environments available. Above all, CODESYS supports all the primary IEC 61131-3 languages and offers support for many of the advanced IEC 61131-3 features, such as object-oriented programming.

It may seem almost oxymoronic because if you ask a PLC programmer what the most powerful PLC programming suite is, they will probably answer with something along the lines of S7 or RSLOGIX. However, in terms of unbridled power, CODESYS and the systems that derive from CODESYS, such as Beckhoff's TwinCAT, offer significantly more features that, when used to the fullest, allow developers to produce vastly superior software. CODESYS and similar programming suites could very easily be the way of the future, and if you can master the features in CODESYS, you will be able to pick up a programming software such as RSLOGIX with ease.

The following are some of the major attributes of CODESYS that the average developer would recognize:

- Full support for object-oriented programming
- A built-in simulator that allows code to be run without physical hardware
- A built-in debugger
- Support for all IEC 61131-3 programming languages

These are just some general, high-level features; some of these features are available in other PLC programming systems as well. For the inexperienced, these need to be highlighted, as some of them will be used quite often throughout this book.

Now, this book will use CODESYS as the programming environment. However, it should be noted that this book is not a CODESYS book. The techniques demonstrated here are simply programming concepts that can be used in any other IEC 61131-3 programming environment or system that closely follows the standard's syntax. With that, you can opt to use a system such as Beckhoff's TwinCAT if you wish. TwinCAT and similar systems such as WAGO and PLCnext are built on top of CODESYS and are IEC 61131-3-compliant, which means that what we cover here is compatible with those systems.

To follow along with this book, you're going to use a working copy of CODESYS. If you've used something akin to TwinCAT, you can use that if you're comfortable, but if you're inexperienced and want to follow along with the book, it is best to use CODESYS. The next section is dedicated to installing CODESYS.

Installing CODESYS

Obviously, before you can move forward with the book, you will need a working copy of CODESYS. CODESYS is free to download and install. The link to download the programming system can be found in the *Technical requirements* section of the chapter. The first step is to follow the link and install the programming environment.

Before you start trying to install CODESYS, it is important to realize that most software, no matter the vendor or purpose, will be periodically updated and changed. This means that, depending on when you're reading this book, there may be changes to the installation process or operation of the software. Typically, these changes are minor; however, beware that as time progresses, these changes may cause incongruities with the instructions presented here. If you do find incongruities, usually a simple Google search will reveal the correct course of action.

> **Note**
> CODESYS versions are not always backward-compatible. If you are using a newer version of CODESYS and the examples do not run, try copying and pasting the source code into a new project built with the version you are using.

To install the program, all you have to do is download the software and follow the wizard. There are no gotchas when installing the program. However, you may need to set up an account, which just requires you to provide an email address and password, along with some other information. In all, the process should take about 15 minutes from start to finish.

A CODESYS exploration

Once CODESYS is installed and you start the program, you should be met with something like *Figure 7.9*:

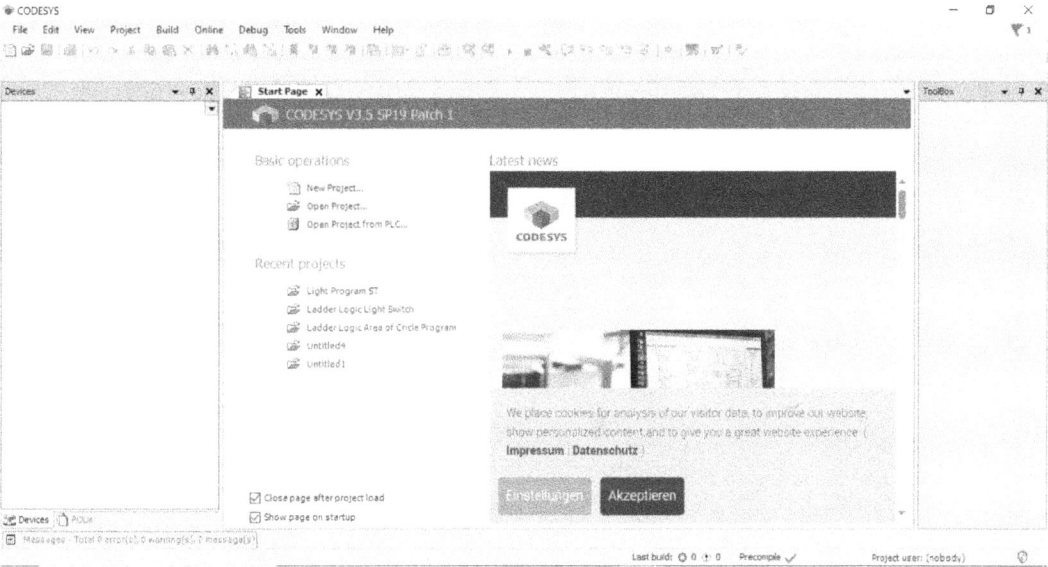

Figure 7.9 – The CODESYS home screen

Figure 7.9 is essentially the landing page for the PLC programming system. To create a new project, click the **New Project…** link, which should render *Figure 7.10*:

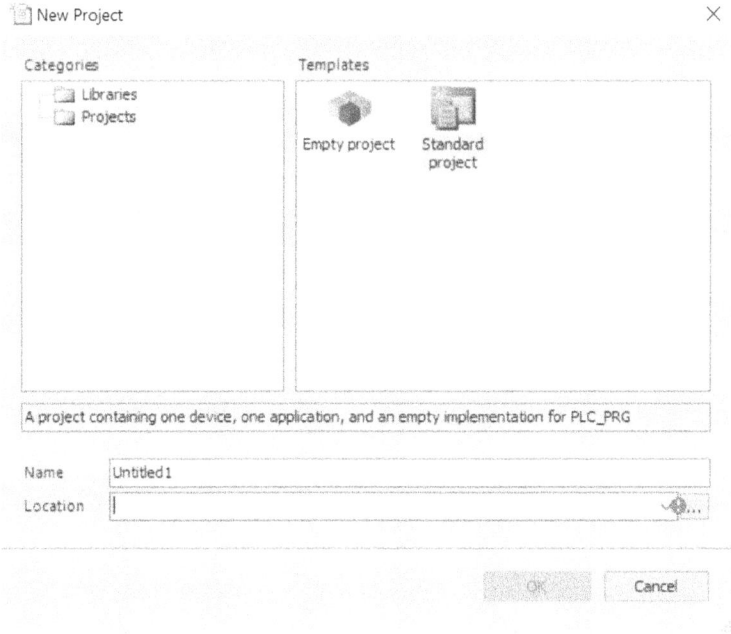

Figure 7.10 – The New Project window

From here, select a location where you want to create your project. The default directory is fine. Input a name in the **Name** field that reflects the nature of the project. When you click **OK**, you should be met with *Figure 7.11*:

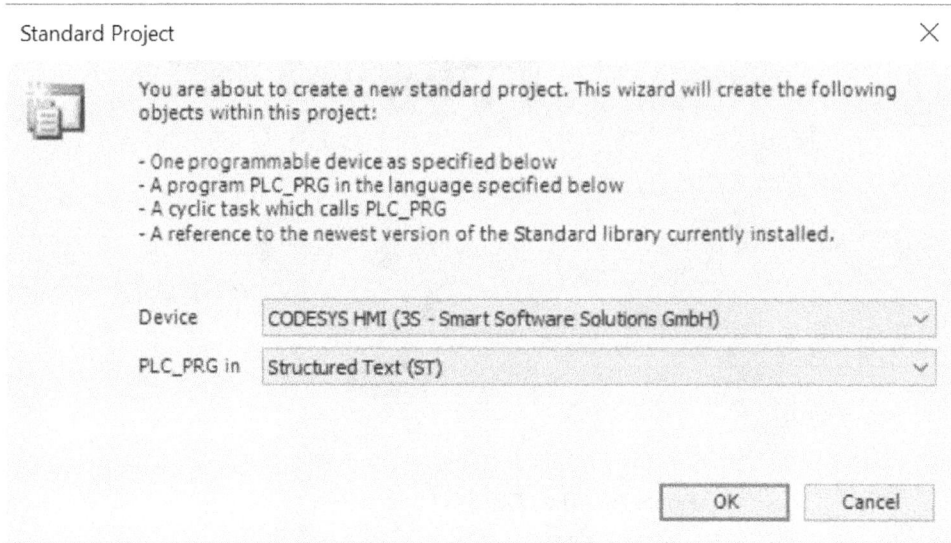

Figure 7.11 – The Standard Project window

Figure 7.11 is a very important window; this window is used to select project properties such as the programming language for the project. By default, ST should be preselected; if not, select it in the drop-down menu and click **OK**.

Typically, from here, you will select a standard project that will create the barebones skeleton of a PLC project. Then, a window like the one in *Figure 7.12* should pop up. In the case of *Figure 7.12*, the PLC_PRG file has been opened. To do this, simply double-click that file in the left-hand tree:

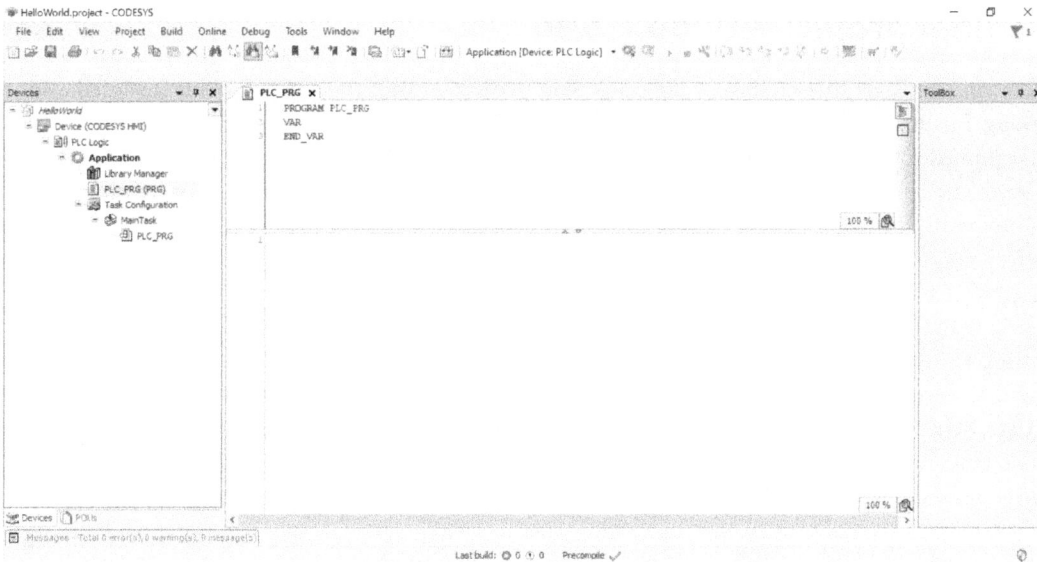

Figure 7.12 – The main project window with the PLC_PRG file open

PLC programs are often broken out among many different files. For example, a PLC program can be broken out among files such as function block files and function files. For this book, we're going to only write code in the PLC_PRG file to keep things simple.

Exploring the PLC_PRG file

Each program file in most CODESYS-based systems is split between a variable section and a logic section. The variable section is at the top, and this is where variables are declared. In contrast, the bottom section is where the core logic of the program goes. Code such as the following goes in the bottom section of the window:

- Math computations
- Assignments
- Control statements
- Function calls
- Function block calls

Once you have CODESYS installed and you feel comfortable with the PLC_PRG file structure, you can move on to the next section – writing your first PLC program!

The final project – Hello World

Before you can start this section of the book, you need to ensure that CODESYS is installed and working. Ensure you can create a new project as outlined in the previous steps. If you can't do that, you will need to go back and fix whatever issue caused the problem with the program installation. Once you are sure you have a working copy of CODESYS, you can start the process of creating your first program.

Step 1 – Creating a new project!

Creating a project was walked through in the previous section. You essentially want to follow the same steps that were outlined previously. Once you have clicked **Create New project**, name the project Hello World, and then click **OK**. Then, click the **OK** button again on the following popup. Once you do that, your project should be generated. Click on the PLC_PRG file in the left-hand tree, and you should see a window like the one in *Figure 7.12*. When you complete those steps, you can move on to *step 2* and start implementing your code:

Step 2 – Code implementation

For this program, we're only going to need one variable called msg. You will want to declare the variable between the VAR and END_VAR keywords in the top part of the window. Once complete, the variable section of your file should match the following:

```
PROGRAM PLC_PRG
VAR
    msg : WSTRING;
END_VAR
```

Now, what this code block does is allocate a memory block called msg and assign it a datatype of WSTRING. Again, don't worry if this does not make sense at this stage; just follow along.

Once your variable is declared, you can implement your logic, which will consist of the following line:

```
msg := "Hello world";
```

All this line does is assign the Hello World message to the msg variable. So, when the program is running, "Hello World" appears in the Value column.

When all this is complete, you can move on to running the program!

Step 3 – Running the program!

For this step, the first thing you are going to want to do is put CODESYS in simulation mode. This setting will run the code locally as opposed to looking for an attached PLC. It is very easy to forget this step. If this step is forgotten, CODESYS will return an error message if there is no physical PLC attached. To do this, you will want to find **Online** in the top menu bar and click **Simulation**.

Once you have Simulation selected, you will need to log in. To do this, click on the highlighted button in *Figure 7.13*:

Figure 7.13 – The login button

Once logged in, click the **OK** button on the popup that appears and then the play button next to the login button, which is highlighted in *Figure 7.14*:

Figure 7.14 – The play button

After play is pressed, your program screen should morph into *Figure 7.15*:

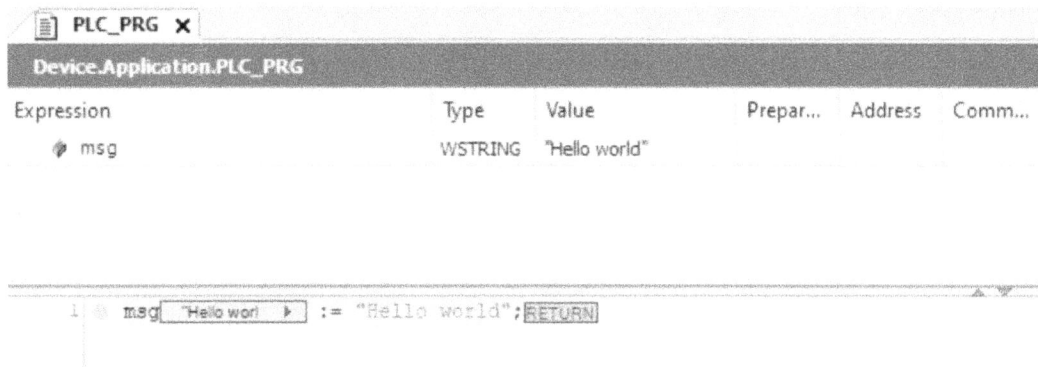

Figure 7.15 – Output

Note the message in the Value column. After the program was run, it reflected the message that was set in *step 2*. After logging in and running, if you see this message, your program has executed successfully!

Chapter challenge

Once you have the final project working, experiment with the message. As a first challenge, try to replace the Hello World message with your name. Once you've done that, add a second variable of type WSTRING, such as the msg variable, and set it to your birthday. Finally, add a third WSTRING variable, set it to your place of birth, and run the program. When you run your program, see what each variable in the Value column says.

Summary

In this chapter, we explored the basics of ST and why it is important, CODESYS, and much more! Hopefully, any fears you had about programming in ST are now fading. At this point, you should have a very general understanding of what CODESYS is, what ST is, and, above all else, the confidence to continue learning the ins and outs of PLC programming in ST. In the next chapter, we're going to take what we learned here and apply it to variables!

Further reading

Basics of ST (Structured Text) Programming: `https://realpars.com/structured-text/`

Questions

1. What are two programming languages that are similar to ST?
2. What are the two sections of the `PLC_PRG` file?
3. Why would you use ST over LL?
4. What does the `Value` column represent?
5. What is the process to run a PLC program in CODESYS?

8
Exploring Variables and Tags

In *Chapter 4*, we explored the basics of how computer memory works. In this chapter, we're going to take those principles and apply them to PLC programming. Unless you're working on a very trivial program, you're going to have to implement variables at one point or another. Moreover, it is a necessity for a programmer, no matter whether they are a PLC programmer or a person writing code for more traditional applications, to understand what variables are and how they work.

In automation, variables are everything. If a sensor is connected to a PLC, the software will require a variable to read the incoming data. If the PLC needs to take an input, a variable will be needed to store the input data. All in all, variables are the world in automation programming. No matter what you're doing, you're going to need to understand variables and how to use them.

Variables are the backbone of any program, no matter the application. This means that to be able to write a program, you will need a solid understanding of how variables work. This chapter is going to be another applied chapter that will explore how to use variables with practical examples. To do this, we're going to explore the following concepts:

- What is a variable/tag?
- How do variables/tags work under the hood?
- Data types
- How to declare a variable
- Variable naming conventions

To round out this chapter, we're going to create a series of variables for a PLC program that can calculate the area of a triangle. By the end of this chapter, you should gain a thorough understanding of data types, variables, and their applications.

Technical requirements

For this chapter, all that will be required is a working copy of CODESYS. All code will be available at the following link:

`https://github.com/PacktPublishing/PLCs-for-Beginners`

It is recommended that you pull down the code and explore the examples.

What are variables/tags?

If you recall from *Chapter 4*, computer – or in this case, PLC – memory is made of storage blocks that can have very long and complex naming conventions that are usually unreadable to humans. To remedy this, programmers will use what are known as variables or tags to provide an alias and context to memory blocks.

A **variable** in programming is much like a variable in mathematics. Ultimately, a variable is a placeholder for a value that may or may not be known. The only area where a programming variable will differ from a math variable is that a programming variable represents a place or location as opposed to a value, as in mathematics. This stems from its nature as an alias.

In traditional programming, these aliases are always called variables. However, in automation programming, these aliases are more often called tags, especially when working with more traditional PLC programming software. Some programmers may split hairs about the difference between a tag and a variable; however, in practice, the terms can be used interchangeably. For this book, we will use the terms interchangeably.

Applications of variables

Variables have many applications. Some common applications that variables are used for are as follows:

- **Internal data manipulation**: Many times, variables are used to hold things such as calculated values, states that were modified by the program, and so on. For programmers who are familiar with Ladder Logic, this concept can be thought of as an internal contact.

- **Input and output**: Variables are often tied to devices such as sensors. Essentially, something such as a pressure sensor will be wired to an input module. Typically, in the program, the value that is read by the PLC will be stored in a variable. On the inverse to that, a variable is typically assigned to an output port. When a value is written to the variable, it will reflect in the physical world.

- **Operator input**: All buttons, inputs, and so on that are in an HMI will be attached to a variable of some kind under the hood. Any control in an HMI must be attached to a variable as each action will require a value to be mutated. For example, a button press may change a value from `True` to `False` or vice versa, a knob may change a numerical value, and so on.

Now that we know a little about what variables are, we need to understand how they work under the hood. Luckily, we explored some of this in *Chapter 4*, so ensure you read that chapter before proceeding. With that, let's explore what variables are under the hood!

Variables/tags under the hood

This section will be relatively short as we've already explored how digital memory works. In *Chapter 4*, we saw that a memory block is an alphanumeric string along the lines of the following:

0x01ABC223

It may not be the same, but this would be along the lines of what a memory address would look like. For us mere mortals, this string has no meaning; chances are, 10 minutes after allocating the memory block, we're going to forget what the memory block was allocated for. This block has no context, so another programmer would have no clue what purpose it serves in a program, and good luck troubleshooting errors.

Why use variables?

A variable solves these problem. As was discussed in *Chapter 4*, a variable is a human-readable alias that references a memory block. Since a variable is a human-readable alias, it provides the following benefits:

- Gives a memory block an easy-to-read name
- Provides context to what the memory block does
- Makes the program more maintainable
- Provides an easy way to change multiple values at once

Now, this section was just a high-level explanation of how a variable works under the hood. However, creating a variable is a little more complex. In the next section, we're going to look at a fundamental of variables, which is called typing.

Data types

Understanding data types is a very important concept for any programmer. To properly leverage any PLC programming language, a strong knowledge of data typing is essential. Typing is a generic concept that is used in every programming language, including languages such as Ladder Logic. In this section, we're going to explore what data types are and how to use them.

What is a data type?

Whether it be a computer, PLC, or any other programmable device, if you need to create a variable, the program will need to know what type of data that memory block can hold. In other words, the program will need to know whether a variable is holding a whole number, a Boolean value, a string, or whatever else it may be. This is where data types come into play. A data type dictates what values can and cannot be stored in the variable.

There are some rules that govern typing, such as whether a programming language is strongly or weakly typed while also being either dynamically or statically typed. This means that a language will fall somewhere in *Figure 8.1*:

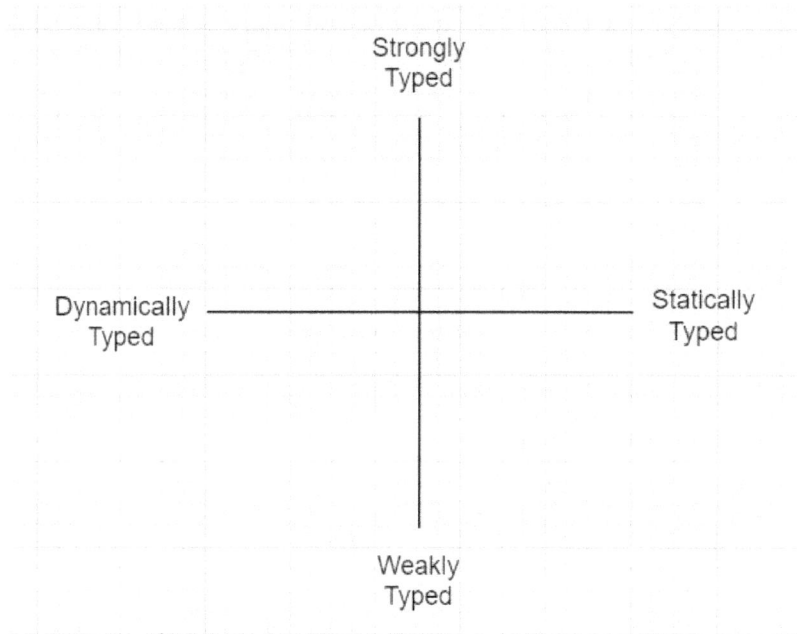

Figure 8.1 – Typing categories

In theory, any language will fall into one of the quadrants in *Figure 8.1*. This means that a language can be one of the following:

- Strongly and statically typed
- Strongly and dynamically typed
- Weakly and statically typed
- Weakly and dynamically typed

For all practical purposes, the definitions of these terms are at best muddy, especially the true meaning of weakly and strongly typed languages. As such, the next two sections are going to explore the concepts at a high level. There is a lot to these terms, but we're going to use the following definitions for this book.

Statically and dynamically typed languages

Describing the difference between a statically and dynamically typed language is a common interview question for traditional app developers and even some PLC programmers interviewing at hi-tech firms. In short, the difference can be summarized with the following:

- **Statically typed**: For practical purposes, in a statically typed language, a programmer will usually explicitly declare a variable's data type when the variable is declared. In other words, a data type is explicitly declared before the compilation process. This means that the type is checked during the compilation process. Common examples of statically typed languages are C#, Java, and C++.

- **Dynamically typed**: In a dynamically typed language, the programmer usually does not explicitly declare the data type and the compilation system figures out the data type based on the value assigned to it. In the case of a dynamically typed language, a variable's data type is determined during runtime. Common examples of dynamically typed languages are Python, PHP, JavaScript, List, Ruby, Objective-C, and so on.

Whether a language is statically or dynamically typed poses its own set of challenges. For example, a statically typed language requires more upfront thought and design work regarding the type of data that a variable should hold. On the other hand, it can sometimes be challenging to figure out a variable's data type in a dynamically typed language. The ambiguity can often cause bugs in a program and make troubleshooting difficult. With that, the rules for a weakly and strongly typed language are less clear.

Weakly and strongly typed languages

This is where things can get murky, quick. The definitions of a strongly and weakly typed language and how they relate to dynamically and weakly typed languages are not agreed upon and can differ depending on whom you ask. For this book, we're going to define the language types as the following:

- **Strongly typed**: A strongly typed language is a language that strictly enforces typing rules. In other words, in a strongly typed language, this line of code would result in a compilation failure:

```
123 = "123"
```

For this line of code, the compilation system will attempt to compare an integer to a string type. For strongly typed systems, the system will only allow operations to be performed on compatible data types. Typically, bugs can be troubleshot more easily in a strongly typed language due to the strict typing rules. Moreover, bugs will be easier to find as the compilation system will check for data mismatches and flag those errors during compilation. Common examples of strongly typed languages are Python, C++, C#, Java, and the like.

- **Weakly typed**: A weakly typed language is the opposite of a strongly typed language. In a weakly typed language, data types don't matter as much. By that, a program will attempt to evaluate the statement. Depending on the language, it will attempt to convert one of the values from a string to an integer or vice versa and conduct the comparison. Obviously, this could cause bugs in a program, as depending on what the program is doing, they may not know whether the program is testing a string comparison or a numerical value. Now, there is nothing wrong with using a weakly typed language; in fact, one of the most widely used programming languages, JavaScript, is weakly typed. The only drawback to using a weakly typed language is that programs written in the language can sometimes be difficult to troubleshoot. With that, common examples of weakly typed languages are JavaScript, PHP, Pearl, Ruby, and the like.

Defining weakly or strongly typed languages is a very ambiguous task that, again, stems from how loosely defined the two concepts are. There is also the "issue" that many modern languages such as C# have adopted many features that can be considered dynamically, statically, weakly, and strongly typed features. In all, this explanation will generally get you through an interview.

IEC 61131-3 typing

If one examines the previous definitions, CODESYS would be a statically and strongly typed language. This can be shown in *Figure 8.2*:

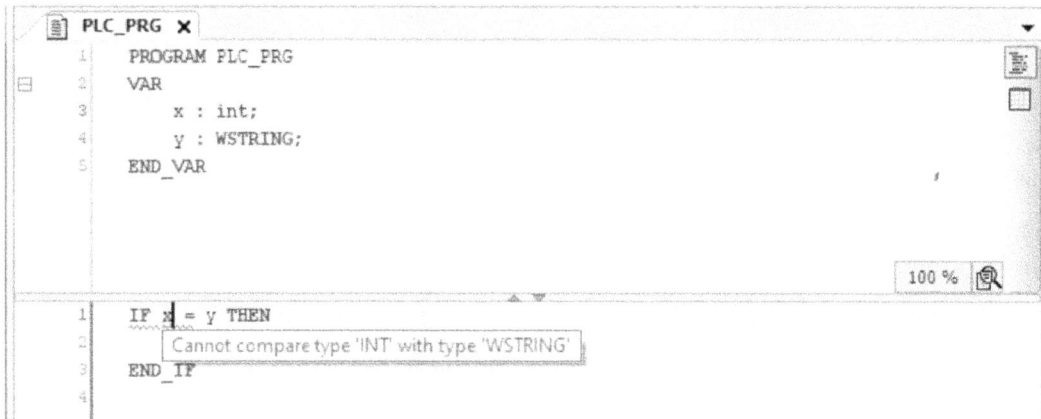

Figure 8.2 – Typing error

As can be seen, there is a red squiggle line in the logic block. Also notice that in the variable block, we have a variable of type INT and of type WSTRING. Now, red squiggle lines typically mean errors and if one hovered over the red line, the message will show that there is an issue with the data types. Since we have to explicitly declare a variable's data type and the system cares about the variable's types, we can say that we are working with a statically and strongly typed language.

So, what does this mean? Why should we care? In short, this means we need to take a deep dive into data types and what they mean.

The IEC 61131-3 data types

The IEC 61131-3 standard supports a bunch of different data types. You'll find that as a PLC programmer, you'll use some data types more than others. However, it is important to understand the data types that the standard supports. With that, the first thing we need to do is establish what a bit and a byte are.

Bits and bytes

It is almost impossible to get through a day without hearing the term bit or byte, especially if you work in **information technology** (**IT**) or automation. However, what do they mean and what do they represent?

Bits

A bit is very easy to understand. A bit can be either a 0 or a 1. In other words, a bit can be thought of as a switch. Much like a switch, a bit can be either on or off and nothing else. Bits are the fundamental backbone of computer science and by extension, automation programming. Bits can be thought of as protons and neutrons, the fundamental particles that make up an atom, because when you have enough bits, they form what's called a byte.

Bytes

If bits are analogous to protons and neutrons, a byte can be thought of as an atom. A byte is typically composed of eight bits for a standard system. The combination of the bits that comprise the byte will determine what the byte is. This is a very important concept to understand, as most of the data types that we're about to explore will use bytes as the metric for the size of the value it can hold.

Representation

Anyone who has ever purchased high-speed internet has seen something akin to 60Mbps. For the uninitiated, this would mean 60 megabytes per second. However, there is a gotcha here. Bytes and bits are denoted with an uppercase and lowercase b respectively. This means that if you've ever bought high-speed internet or data of some kind and it was denoted with a lowercase b and you thought or were told you bought megabytes, your speed is actually one-eighth the speed or less!

The denotation of bits and bytes is very important because sometimes, documentation can use either metric. Many communication protocols also use the metrics interchangeably. As such, when working with variables and you need to know how large of a value you can store in the variable, you must know how the units are represented, especially when you're integrating PLCs with general-purpose programming languages such as C# or Java.

Why are bytes and bits important?

This may seem trivial; however, as technology progresses, traditional programming languages such as C/C++, C#, Java, Python, Delphi, JavaScript, and so on are finding their way into the automation world. For example, Beckhoff offers the TwinCat system that allows developers to write HMIs in a .NET or Delphi language. In cases like this, what constitutes an integer on the PLC side is equivalent to a short int or 16-bit integer in a .NET language and is sometimes denoted as such. This means that when the HMI sends data to the PLC or vice versa, there could be data compatibility issues if the data types are not declared correctly. As technology progresses even further and newer technologies such as the cloud, machine learning, and IoT become more prevalent, these traditional, general-purpose programming languages will become more ingrained in the automation world. As such, it will become ever more important to understand how many bytes or bits represent a data type and the units each system uses.

Now that we have a little background about how data types work and the difference between bytes and bits, we can move on to some practical applications for variables. To do this, we are going to explore the data types in a very practical sense. To start the journey, we're going to explore some of the most common data types that a programmer will use.

Common data types

In programming, some data types are used way more than others. In practice, most programmers will find themselves using the following data types the most:

- **Integer**: An integer is denoted by the INT keyword in most languages as well as the IEC 61131-3 standard. This data type signals that the variable will hold a single whole number. An example of an INT is a value such as 1, 23, 1001, or so on. You will typically use an INT for logic such as counting parts, inputting the number of parts to make, and similar things. Essentially, any logic that will handle whole numbers will usually utilize the INT data type.

- **Floating point**: A floating-point data type is a type that holds decimal values. Essentially, any value that has a decimal is typically considered a float. Unlike integers, which are typically denoted by the INT keyword or a derivative of the three letters, a floating point is declared with different keywords depending on the language. Common keywords that denote a floating point are as follows:

  - Float
  - Double
  - Real

This data type is used to hold values that are the result of division, analog inputs, or outputs from devices such as pressure sensors, temperature sensors, scales, and so on. In an IEC 61131-3-compliant language such as CODESYS or Beckhoff, use the REAL keyword to denote a floating point and an LREAL keyword to denote an 8-byte floating point.

- **Boolean**: A Boolean value can either be a 0 or a 1 and nothing else. Basically, a Boolean value is a bit. Boolean variables are typically expressed with the keyword bool or Boolean. As was explored in *Chapter 6*, bools are mostly used to determine the output of something based on the input or vice versa. A common example of a Boolean variable may be to read or write a switch attached to a digital I/O module, a light curtain, IR sensors, LED torches, buzzers, and so on. The main use case for a Boolean variable is in a control statement such as an if statement. In IEC 61131-3, a bool variable is denoted with the BOOL keyword.

- **Char/string**: Chars and strings are two different types of data but are very similar in nature to the inexperienced. A char type can be thought of as a single alphanumerical character such as a letter, number, or symbol whereas a string is one or more alphanumerical characters. For example, the letter a and the symbol ! are char types where abc123@ is a string. Some languages such as Python don't support a char type per se, but they support a string type. However, most languages such as C# or Java treat strings like an array of characters under the hood. Where an array can be thought of as one variable that can hold multiple values. Typically, it is more common to use a string than a char but, again, that depends on the language. In terms of PLC programming, it is common to use strings over char and they are typically employed to do things such as naming files, reading error strings, debugging, and so on. On the other hand, chars are common for things such as taking simple inputs, such as y/n for yes or no. In IEC 61131-3, a string is denoted with the STRING or WSTRING keyword and a char with CHAR.

Of all the data types, these are the ones that you must know to function as a PLC programmer. Many of the other data types that IEC 61131-3 supports are derivatives of these types. These other types include data types such as **long INT (LINT)**, **short INT (SINT)**, and so on. There are many more data types like these to explore, so much so that a whole book could probably be dedicated to just those. The Wikipedia page on the IEC 61131-3 data types is an excellent resource to explore.

If you wish to explore more about the different data types, you can visit this page on Wikipedia:

https://en.wikipedia.org/wiki/IEC_61131-3

The variables that were explored are the ones we will use the most in this book. When other data types are needed, they will be explored at that time.

As such, for now, we're going to switch gears and explore how to declare variables.

How to declare variables

In previous chapters, we've declared variables just enough to limp through our examples. However, we have not explored their syntax in any real depth. At this point, you may have a rough idea of how to declare a variable but in this section, we're going to do a deep dive into how to declare and use variables.

Variable section of a file

All files in CODESYS and similar systems have a special area where variables are declared. As you may have noticed, a file, such as the PLC_PRG file, is split in two. The bottom section is where the program's logic is declared, as we have seen, while the top section is used to define variables.

Any time you create a file, the top section will always have the following lines of code in it by default:

```
VAR
END_VAR
```

This is a special block of code, as this block of code is where all the variables for that particular file will be declared.

Declaring a variable

On its own, this block of code doesn't do anything since there is no code in between the two keywords. To add life to this code block, we, at the minimum, need to declare a variable. Declaring a variable is basically just telling the PLC to create a new variable of whatever type we need. The exact syntax for declaring a variable is as follows:

```
VAR
     Variable_name : data type;
END_VAR
```

So, if you needed to declare a variable called counter of type INT, you would use the following:

```
VAR
     counter : INT;
END_VAR
```

This code will declare a variable called counter of type INT, but that value will be empty. For some cases, this is fine, but for others, you may want to preload a value into the variable.

Initializing a variable

Often, we will want to assign a value to a variable when it is declared. This is especially true if you're working with a dynamically typed language such as Python, as a language like that will need to know the data type when the variable is declared. For our purposes, we can technically skimp on this since we are statically telling the PLC what type of data that memory block is going to hold. Though it is not necessary, it is considered a best practice by some to assign a dummy value such as 0 to a variable at declaration, just so it is not empty.

An example of this may be an industrial oven. For industrial ovens, it is common to have a preloaded cutoff temperature that will serve as the default cutoff temperature for most of the parts that go into it. For the most part, a programmer should not have to worry about coding this value in the logic. In fact, that would be a bad idea. In practice, they should just declare the variable and assign a value to it all-in-one shot. To do this, they could use the following syntax:

```
VAR
      oven_temp : INT := 320;
END_VAR
```

The key here is the := sign. In the IEC 61131-3 standard, that syntax means assign. In other words, drop a value into the variable. As can be seen in the syntax, we are declaring a variable in a normal fashion, but we are assigning a value to it at the same time. In this case, we are initializing the variable at declaration. As such, anywhere oven_temp is referenced, the value will be 320 unless that value is changed somewhere else in the logic.

Variable naming

Believe it or not, one of the keys to a well-written program is the naming conventions of variables. If you think back to our apartment example in *Chapter 4*, how many times have you said that you're going to go visit the residents in apartment 123 at 456 Jay Street? Chances are, you've never said that. On the contrary, you've probably said you're going to go visit your friends Bob and Sarah. Memorizing Bob and Sarah's address is simply too difficult to do and keep track of, especially when you are a social butterfly and have thousands of friends. Programming is no different. A variable's name adds context to what the value is and its purpose in the program as such, so to keep your code base clean and concise, effective naming is a must! So, in this section, we're going to look at how to properly name variables and proper naming conventions.

Rules to naming a variable

Before we start exploring proper naming conventions, we need a little computer history lesson. Up until recently, and even today when using some old PLCs, computer memory was precious. Computers and PLCs alike did not have the resources necessary for long variable names. It was common to have code bases with X and Y as variable names. These names are technically better than a raw memory address but they're not very helpful in understanding what the variable is and its purpose in a program, especially when the program has thousands of variables. Fast forward to today, and computer and PLC memory isn't as precious anymore and variable names that add context are the new norm.

In today's landscape, a variable has three main tasks:

- Add context to the memory address's purpose
- Provide a logical, human-readable alias for a memory block
- Store a value in a central location

This means that naming a variable X or Y isn't a good practice anymore.

A good rule of thumb is to name a variable based on what it is meant to do. For example, suppose you want to use a variable to set the speed of a motor and suppose a device has three motors. Logical names for the motors would be as follows:

- motor1Speed
- motor2Speed
- motor3Speed

Or

- speedMotor1
- speedMotor2
- speedMotor3

As can be seen, each of the variables clearly indicates what the variable does and what device it controls. So, if another programmer or your future self is examining the code, the responsibility of the variable is clear.

Now, there is no silver bullet to naming variables; however, a good pattern to follow is to include the device, responsibility, and device number if applicable. This pattern will generally serve you well; however, sometimes you'll name a variable something like greenLED. Depending on what you're working on, you could probably get away with this. However, a better name would be something along the lines of startLEDOn. All in all, learning to name a variable is a bit of an art and a science.

With that, the next thing we need to look at is naming conventions.

Naming conventions

Naming conventions are more than just naming a variable. Proper naming conventions also include tasks such as properly formatting the variable's name. There are three main conventions that are commonly used to do this.

Camel casing

Arguably the most common naming convention is what's called **camel casing**. As was explored in the last section, many variable names are composed of multiple words. In camel casing, the first word is spelled using all lowercase while the first letter of each subsequent word is capitalized. This means that if someone wanted to declare a speed variable for motor 1, they would use the following:

```
motor1Speed : <data type>
```

Camel casing is very common, but it is not the only naming convention. Another common naming convention is pascal casing.

Pascal casing

Another popular schema is to use what is known as **pascal casing**. For this schema, the first letter of each word in the variable is capitalized. Circling back to the motor speed variable, in pascal casing, the variable would be written as follows:

```
Motor1Speed : <data type>
```

Pascal casing and camel casing are both very common and easy-to-use naming conventions. However, there is one more naming convention that is commonly used called snake casing.

Snake casing

Snake casing is an excellent convention to use but it can be difficult to use compared to the other two. In snake case, each word is separated by an underscore, which makes the variable look like a snake! Of all the naming conventions, this one can be the most awkward to use due to the placement of the underscore button on a keyboard. For this convention, every word is typically spelled with all lowercase letters and each word is separated by an underscore. This means that if one wanted to make a motor speed variable with snake case, they would use the following:

```
motor_speed
```

In any case, it doesn't matter which convention you use if it is consistent throughout the program. Usually, an organization will have a coding standard that dictates what schema to use. For this book, we're going to exclusively use camel casing.

We now have the information needed to properly declare variables. Now that we understand data types and naming conventions, we can move on to our final project. For the final project, we're going to declare the variables necessary to calculate the area of a triangle.

Final project – declare the variables of a triangle

In PLC programming, it is common to have to calculate the area of shapes. A common shape that is often seen in automation is the good 'ol triangle. For this project, we're going to set up the necessary variables to calculate the area of a triangle. We're going to pretend that the our program is going to control a pump that will pump cement into triangular containers that are all the same size.

The first thing we need to establish is what the area of a triangle is. The equation to accomplish this task is as follows:

$$Area = \frac{Base * Height}{2}$$

For this program, we're going to make two assumptions. We're going to assume that both the height and the base are whole numbers. We're also going to assume that the base is 2 and the height is 10.

Code implementation

Breaking down the aforementioned scenario list, we know that the height and the base are both whole numbers, so they can be declared as an integer type. Also, for the equation, we know that the calculated area is the resultant of division, so that needs to be a floating-point data type. As such, we can use the following code to implement the skeleton of the variables:

```
VAR
    area    : REAL;
    base    : INT;
    height : INT;
END_VAR
```

Now, there's one more step. Since the scenario stated that the base and height are constant, we can modify the code to accommodate for that. We can initialize the base and height variables with the following:

```
VAR
    area    : REAL;
    base    : INT := 2;
    height : INT := 10;
END_VAR
```

In this case, we preloaded the values, so we don't have to modify the logic to do this. All in all, our variables are declared and satisfy the requirements of the scenario.

Challenge – declare the variable for a rectangle's perimeter

As a special challenge, assume the perimeter of a rectangle is defined with the following equation:

$$Perimeter = 2(length + width)$$

For this challenge, assume that the length and the width can be either a whole number or a decimal number. Declare all the variables for this program.

Summary

In summary, we've explored the basics of typing, common data types, variables, and more in this chapter. By this point, you should be able to effectively declare and name a variable to create quality code. Now, don't be discouraged if you don't fully understand the ins and outs of variables yet, because typing and naming conventions can often be confusing to the inexperienced. You'll get much more practice in the next chapter, when we explore math calculations in Structured Text!

Further reading

- *A guide to common naming conventions*:

  ```
  https://www.theserverside.com/feature/A-guide-to-common-
  variable-naming-conventions#:~:text=The%20standard%20naming%20
  conventions%20used,snake%20case
  ```

- *Static and Dynamic typing? Strong and weak typing?*:

  ```
  https://dev.to/leolas95/static-and-dynamic-typing-strong-and-
  weak-typing-5b0m
  ```

Questions

1. What data type is used to store whole numbers?

2. What is a floating-point number?

3. What data type is used to store a floating-point number?

4. What data type would you use to store the result of 3/2?

5. What is a strongly typed language?

6. What is a data type?

7. What is a weakly typed language?

8. What is a dynamically typed language?

9. What is a statically typed language?

10. Is CODESYS dynamically or statically typed?

9

Performing Calculations in Structured Text

Math – the nightmare fuel for many students around the world and the one subject that everyone loves to hate. Unfortunately, to be successful as a PLC programmer, you must understand how to program mathematical equations. Luckily, performing math for PLC programs does not require any calculations that must be done by hand. This means that you don't need a math degree to be a great PLC programmer.

As an automation engineer, you'll often find yourself working on machines that require complex mathematics. Machines that must calculate fill rates, pressure, and of course motion control will all require the program to handle complex math. Put bluntly, math is a major area of PLC programming that all programmers must be competent in.

This chapter is going to focus on crunching numbers in **Structured Text** (**ST**). Generally, doing complex calculations such as those that are required for proper motion control, fill rates, and so on can be quite complex, awkward, and, in many cases, nearly impossible to effectively program in **Ladder Logic** (**LL**). In short, when it comes to math that is more complex than basic four-function calculations and relatively simple algebraic equations, ST is vastly superior in terms of writability, understandability, and maintainability. However, those not well-versed in ST can find it daunting.

To explore mathematics in ST, we're going to cover the following concepts:

- Math in ST
- Assignments
- Basic mathematical functions in ST
- Complex mathematical functions in ST
- Trigonometry functions in ST

- Order of operation in ST

- Complex equations in ST

Unlike the previous chapter in this book, we're going to do two final projects. The first project to get our feet wet will be the solution to the previous chapter's project challenge; for this, we will calculate the perimeter of a rectangle. To round out this chapter, we're going to write a PLC program to calculate the hypotenuse of a triangle using the Pythagorean theorem.

Technical requirements

For this chapter, all that will be required is a working copy of CODESYS. As such, if you have not already done so, you will need to complete the setup presented in *Chapter 7*. Once you are done with that, you will be able to follow along with this chapter. All the code examples are available at `https://github.com/PacktPublishing/PLCs-for-Beginners`.

To understand this chapter, you will need to understand the data types that were explored in the previous chapter. As such, if you do not understand that material or haven't at least read the material in the previous chapter, you need to go back to that chapter.

Math in ST

Math and automation go hand in hand. Most machinery will need to crunch numbers at some point to safely and successfully carry out its tasks. However, as we saw in *Chapter 7*, calculating numbers in LL can be rather challenging and awkward, especially for large, complex equations that are synonymous with tasks such as motion control and precision operations.

As we also saw in *Chapter 7*, math in ST is very intuitive and more closely resembles what we all grew up with. Equations are written in an easy-to-read and understandable format using common mathematical symbols. Though it can seem intimidating at first, math in ST is very straightforward and can save you lots of development time in the long run, especially when you must compute large complex equations. In all, ST helps in computing complex equations as it allows you to do the following:

- Write equations naturally.

- Program equations in an easy-to-read format.

- Quickly troubleshoot complex equations.

To begin our exploration of math in ST, we'll start by looking at one of the core basics of math: assigning numbers.

Assigning numbers

The first step in calculating numbers is assigning values to variables. In the previous chapter, we touched on the concept of initializing a variable with a value, during declaration, in the variable section of the program file. However, as stated in that chapter, it is possible to assign a value in the logic section. Typically, you'll assign a value to a variable in the logic section of the file when you do a computation.

Unlike most other programming languages, such as Java, C++, or C#, the assignment operator is not the typic equals sign. For the IEC 61131-3 standard, the assignment operator is denoted with the following symbol:

```
:=
```

Essentially, this operator tells the PLC to assign a value to a variable. For example, this block of code will assign the number 3 to the x variable:

```
x := 3;
```

The following block of code will assign the value in the a variable to the b variable:

```
b := a;
```

As we can see there is a pattern for assignments. The value on the left of the assignment operator is the target value, which means this is the value that will change when that line of code is run. The value on the right of the operator is the value that will be placed in the target variable.

With assignments under our belts, we can move on to more interesting programming: numeric programming. With that, we're going to begin exploring numerical programming by covering the four fundamental operations of mathematics: addition, subtraction, multiplication, and division.

Basic calculations

As logic has it, the basics of any calculations are the four basic operations of addition, subtraction, multiplication, and division. If you've worked with LL in the past, you may have noticed that the programming interface utilizes function blocks that have to be strung together, which can make long equations relatively complex. This is where ST shines as equations that are comprised of many variables can be easily programmed.

Solution variable

All operations have solutions – that is, all equations have an output value that is the result of the calculation. In programming, it's usually wise to have an extra variable to hold the resultant of an equation. For example, if you were to add two numbers together, you may have a third variable called sum and your code may look like something akin to the following:

```
sum = a + b
```

Using a solution variable will help keep your code cleaner and easier to maintain. However, there are times when you won't need a resultant variable and you can simply manipulate the output; however, for this chapter, we're going to keep things simple and clean. Therefore, all equations will utilize an extra variable to hold the resultant.

The four basic functions

The easiest place to start exploring mathematics is with the four basic operations we all learned about in grade school. As we all know, the four basic operations of mathematics are as follows:

- Addition, which is denoted with the + symbol

- Subtraction, which is denoted with the – symbol

- Multiplication, which is denoted with the * symbol

- Division, which is denoted with the / symbol

Using these operations is quite easy. The basic pattern for using these operational functions is as follows:

```
result_variable := value1 <symbol> value2
```

This pattern is all you will need to start computing basic equations. Now, let's experiment with the four basic operations.

Basic operation demonstration

To demonstrate the four basic functions, we'll create four variables called sum, difference, product, and quotient. All the variables will be of the INT type, except for the quotient, which will be of the REAL type. Hence, you should set up the variable section of your PLC_PRG file so that it matches the following:

```
PROGRAM PLC_PRG
VAR
    sum             : INT;
    difference      : INT;
```

```
        product          : INT;
        quotient         : REAL;
END_VAR
```

The main logic will perform the following operations:

- Add 3 to 3

- Subtract 2 from 5

- Multiply 6 and 6

- Divide 9 by 3

To do this in ST, you should use the following code:

```
sum          := 3 + 3;
difference := 5 - 2;
product     := 6 * 6;
quotient    := 9 / 3;
```

When the code is executed, you should see what's shown in *Figure 9.1*:

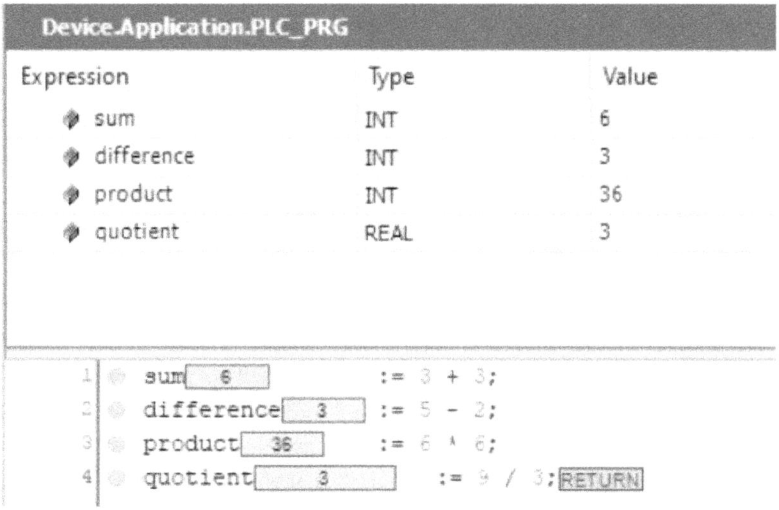

Figure 9.1 – Basic math output

Notice that the outputs are what we would expect if we plugged these values into a calculator.

Real-world calculation example

Now, it is very common to only need to perform a function that requires something such as addition or subtraction. However, you will also encounter problems that will require more complex calculations. For example, suppose you are programming a PLC to control a cement mixing process. Imagine that in the last stage of the operation, the machine has two nozzles that will fill two bags of cement. Once the bags are full, the machine will take the weights of both bags before sending them for palletization. For this machine to work properly, cement must be loaded into the machine's main hopper. For the machine to be topped off, the PLC must send a signal to the hopper when there is less than 20 lbs of cement left in the hopper. The hopper does not have a scale, so the PLC will have to estimate when the hopper needs to be filled again. To do this, we can use the following equation:

$$Current\ Weight\ =\ Amount\ in\ Hopper\ -\ (Bag1\ Weight\ +\ Bag2\ Weight)$$

In the real world, this value would be constantly updated since the batch process would fill more than one bag, as stated in the scenario. For this program, we will need the following variables:

```
PROGRAM PLC_PRG
VAR
     currentHopperWeight            : REAL := 250;
     bag1Weight                     : REAL;
     bag2Weight                     : REAL;
END_VAR
```

In this example, we have three variables. Two variables represent the weight of the bags, while the other represents the current weight of cement in the hopper. Now, this is one of those programs where it makes more sense to reassign the solution to the equation to the hopper weight variable as opposed to making an extra variable. This is because this variable will be a dynamic part of the equation – that is, the variable will change with each set of bags that are set to be filled. Also, in this example, we are initializing the variable to be 250. In the case of this scenario, this will represent a starting weight of 250 lbs.

In terms of the data types, notice that they are all set to REAL. This is very important due to the nature of the application. The weight of the bags can be – and probably will be – decimal values for two reasons. The first obvious reason is that the operator may want bags that are 22.2 lbs or 5.25 lbs. On the other hand, no matter how precise the equipment is, it will never give an exact target weight. For example, if a target weight for a bag is a whole number, such as 45 lbs, the machine will probably overfill or underfill by a certain margin – for example, it may fill the bag to 44.97 lbs or 45.02 lbs. In other words, the machine will have a tolerance, but the PLC will need to know as accurately as possible what each bag's weight is so that it can account for it in the hopper weight equations. If we were to use integers for these numbers, chances are the decimal values would be shaved off and when the equation is running, it would produce an inaccurate weight for the cement in the hopper. This could mean that the bags could be drastically underfilled, or cement that is loaded into the hopper could be wasted. Overall, this is why a clear understanding of data types is vital in PLC programming!

With all that, the logic for the equation can mirror the equation:

```
currentHopperWeight := currentHopperWeight - (bag1Weight +
bag2Weight);
```

In this equation, the `currentHopperWeight` variable is used as a reference point. When the bags are weighed, the sum of the weights is subtracted from `currentHopperWeight` and then that value is reassigned to the `currentHopperWeight` variable.

To test the program, we can run the program and write two values for the bag variables. For the example, we will use 50 for each bag of cement. Enter 50 in the prepared values field in the grid, as shown in *Figure 9.2*:

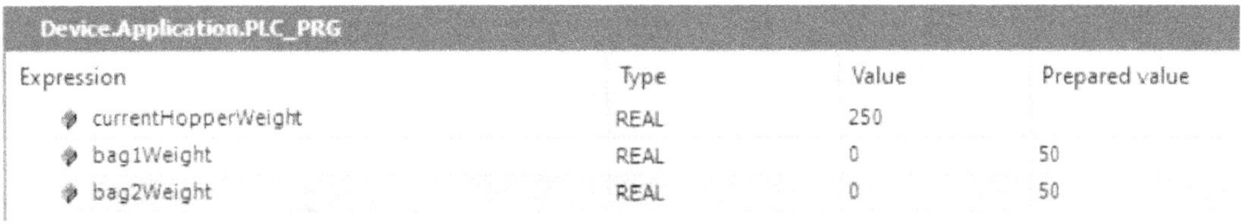

Device.Application.PLC_PRG			
Expression	Type	Value	Prepared value
currentHopperWeight	REAL	250	
bag1Weight	REAL	0	50
bag2Weight	REAL	0	50

Figure 9.2 – Prepared values

Once you have the values in the field, as shown in *Figure 9.2*, you will need to right-click in the grid and press **Write All Values**, as shown in *Figure 9.3*. This will overwrite the current value in the fields – that is, the 0 weight for each bag – with the number 50. Now, watch the `currentHopperWeight` variable continuously tick down until it is a negative number. Essentially, the program will simulate 100 lbs of cement leaving the hopper each time a bag passes.

As the program runs, `currentHopperWeight` will keep getting smaller and smaller. This is not a bug but intended behavior. What's happening is that the ST PLC program will continuously loop the same way an LL PLC program will. This means that even though we have no looping logic, the `currentHopperWeight` variable will get continuously recalculated each cycle and since the bag weights are statically set to 50, each 100 will be subtracted from the original 250 value that resided in `currentHopperWeight`. In a real-world setting, a programmer would utilize sensors and other components to detect a bag, which would prevent runaway calculations:

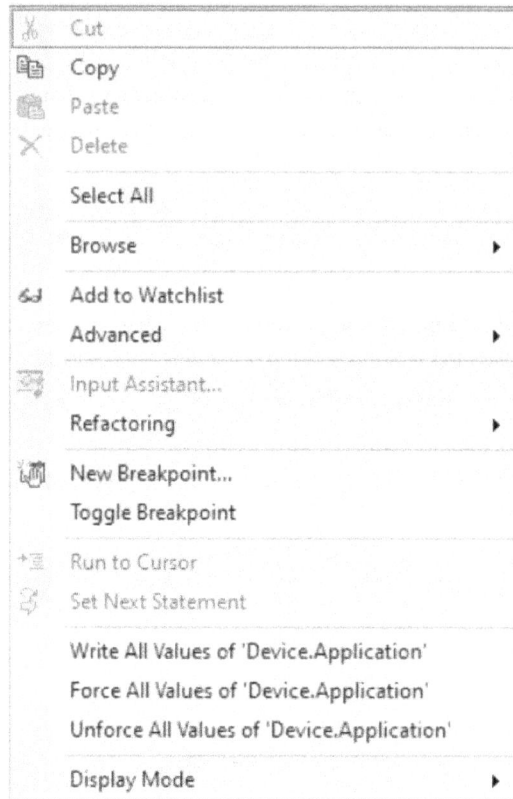

Figure 9.3 – The Write All Values option

So, now that we have some basics of computational programming under our belts, let's look at the advanced mathematical functions.

Complex mathematical functions

Chances are you're going to be doing much more than simple addition, subtraction, multiplication, and division in your PLC programs. Most likely (especially for motion control), you're going to utilize some more advanced mathematical functions, such as square roots, exponents, and trigonometric functions. So, in this section, we're going to explore those functions and how to use them in a PLC program.

Square root function

Outside of addition, subtraction, multiplication, and division, square roots are one of the most common math functions there are. Many basic geometry and trigonometric operations that PLC programmers will use in their day-to-day lives will depend on the square root of a number. The best way to explore this function is to use it in an example.

As usual, the first thing you should do is set up the program variables. As we progress through this section, we're going to add more variables and functions, but for now, set your variable section so that it matches the following code block:

```
PROGRAM PLC_PRG
VAR
    root : REAL;
END_VAR
```

A square root is simply some number squared that will produce the original number. This means that unless the number you feed into the square root function is a perfect square, which it most likely isn't, you will have a floating point as a return value; consequently, you will want to have your output variable be of the REAL type.

In terms of the actual logic, your program body should look like this:

```
root := SQRT(9);
```

The number we want to find the square root of is in between the parentheses of the SQRT function; in this case, the number is 9. When the program is running, your variable grid should look like what's shown in *Figure 9.4*:

Expression	Type	Value
root	REAL	3

Figure 9.4 – Square root output

As can be seen, the output is 3. Now that we've seen the square root in action, let's explore its cousin, the exponent function.

Exponent function

The exponent function works very similarly to the square root function. However, this function will take two parameters – that is, it takes two numbers instead of one as inputs. To understand this function, it is important to first understand the following equation:

$$result = base^{power}$$

This equation will translate into the following code:

```
EXPT(base, power)
```

If we wanted to calculate 4^2, we would need to modify the variable code so that it matches the following:

```
PROGRAM PLC_PRG
VAR
     root   : REAL;
     square : REAL;
END_VAR
```

We will also need to modify the logic so that it matches the following:

```
root := SQRT(9);
square := EXPT(4,2);
```

In this case, we are simply adding to the square root example from before. If you are following along, you can omit the square root code if you wish. When you run the code, you should get the output shown in *Figure 9.5*:

Expression	Type	Value
◆ root	REAL	3
◆ square	REAL	16

Figure 9.5 – Exponent output

As we can see, the square of 4 is 16, just as expected. The next important math function that we will explore is the ABS function.

ABS function

The ABS function will simply return the absolute value of a number. In other words, if a value is negative, it will return a positive value and if the value is positive, it will still return a positive number. We will continue to expand our previous examples and modify the variable section so that it matches the following:

```
PROGRAM PLC_PRG
VAR
     root      : REAL;
     square    : REAL;
     abs_value : REAL;
END_VAR
```

Next, we will modify the logic so that it matches the following:

```
root := SQRT(9);
square := EXPT(4,2);
abs_value := ABS(-3);
```

When the code is running, you should see an output similar to what's shown in *Figure 9.6*:

Expression	Type	Value
◈ root	REAL	3
◈ square	REAL	16
◈ abs_value	REAL	3

Figure 9.6 – Absolute value output

As you can see, the ABS function converted the negative 3 into positive 3, as expected. Now, there are a lot of built-in math functions in the IEC 61131-3 standard. Most programmers will not memorize every function a language or programming technology offers. So, learning to read and understand documentation is key to being a successful programmer. Now, try to solve the following challenge.

Math challenge

Assume you are working on two types of widgets. Suppose Widget A needs its packaging to be rounded down to the smallest whole number and Widget B needs its packaging to be rounded up to the highest whole number. For this challenge, write a simple program that can take numbers for Widget A and Widget B and round the numbers accordingly.

Hint

Do some research online and look for a ceiling and floor function.

Once you figure out the challenge, you are free to move on to the next sections. In the next section, we're going to switch gears and look at some trigonometry functions.

Trigonometric functions

As a PLC programmer, you will often have to use trigonometric principles to accomplish tasks. This means having a solid understanding of how to use trigonometric functions will be vital to your success as an automation engineer and, by extension, a PLC programmer. This section will explore the basic trigonometry functions that are supported by the IEC 61131-3 standard. To begin our exploration, we will look at the tangent function.

> **Note**
>
> For this section, a rudimentary understanding of trigonometry will be required and assumed. If you do not have an understanding of basic trigonometric functions, you should spend a little time brushing up on the SIN, COS, and TAN functions.

Anyone who has taken a high school-level trigonometry class will be familiar with the sine, cosine, and tangent functions. Essentially, these are angle properties that can be derived from the sides of a right triangle. As a PLC programmer, it is often necessary to integrate these functions into our program to find things such as height, the length of a makeshift hypotenuse, or another application. Much like the other complex math functions, the IEC 61131-3 standard supports functionality for these operations. To demonstrate these functions, let's set up a new program with the following variables:

```
PROGRAM PLC_PRG
VAR
     tan_of_angle : REAL;
     sin_of_angle : REAL;
     cos_of_angle : REAL;
END_VAR
```

Next, we will need to implement the following code into the body of the logic file:

```
tan_of_angle := TAN(45);
sin_of_angle := SIN(45);
cos_of_angle := COS(45);
```

In this example, we're going to pass in 45 for all the input values. When the program is run, you should be met with the output shown in *Figure 9.7*:

Device.Application.PLC_PRG		
Expression	Type	Value
⚓ tan_of_angle	REAL	1.61977518
⚓ sin_of_angle	REAL	0.8509035
⚓ cos_of_angle	REAL	0.52532196

Figure 9.7 – Basic trigonometric function outputs

> **Note**
> Now, the thing to remember is that the trigonometric function will return its values in radians. This can be a gotcha as some programmers may assume the return value is in degrees.

Arc functions

The next set of trigonometric functions to explore are arc functions. An arc function is simply the inverse of the functions that we just explored. These will also often be utilized in automation engineering. To utilize the arc function, all you have to do is append A to the front of the trigonometric functions that we just explored. This means the arc functions are as follows:

```
ATAN(input);
ASIN(input);
ACOS(input);
```

The past two sections have been a crash course on how to use built-in math functions. At this point, you should have a decent idea of how to perform basic mathematical operations, as well as perform more complex operations, such as finding the absolute value, square roots, and so on. With that, we can now move on and explore the order in which an equation will be calculated.

Order of operations for math calculations

For anyone who has taken a high school-level algebra course, the term PEMDAS may be familiar. In short, **PEMDAS** stands for **Parentheses, Exponents, Multiplication, Division, Addition, and Subtraction**. What this entails is the order of operation in which an equation is calculated. For example, let's say we were presented with the following equation:

$$2 + 3 * 3 + (4 + 4)^2$$

We would need to solve the equation by performing the following steps:

1. $(4 + 4) = 8$.
2. $8^2 = 64$.
3. $3 * 3 = 9$.
4. $2 + 9 + 64 = 75$.

The same general rules apply to writing a PLC program or a program in any other language.

> **Note**
> You must understand the order of operation as not understanding the order of operation can easily lead to wrong calculations, which, in turn, can lead to adverse and potentially dangerous situations.

As you program equations, you'll naturally get a grasp of math ordering. However, following the PEMDAS hierarchy will serve as a good starting point.

You should now have a decent grasp of how to perform calculations in a PLC program. Now, let's take what we've learned and attempt to compute some complex equations that we may run across in our day-to-day lives as PLC programmers.

Computing complex equations

The term complex is subjective, to say the least. For this book, a complex question will refer to a calculation that requires a mixture of the four basic mathematical operations, as well as a complex math function that was previously explored. The first equation that we're going to program is commonly used in automation programming – that is, calculating the distance between two points.

Distance between two points

Calculating the distance between two points is a very common task in PLC programming. It is often necessary to know how long to power motors and so on to ensure the machine reaches its new destination. To figure out the distance, it is common to use the following equation:

$$Distance = \sqrt{(x_2 - x_1)^2 + (y_2 - y_1)^2}$$

In this equation, we will need to subtract two sets of numbers, square two numbers, sum them, and finally take the square root of that sum. As such, we will need to use the `EXPT` function and the `SQRT` function.

There are two ways that this equation can be programmed. We can program this equation all on one line, or we can break this equation down into multiple lines. For an equation this short, it will be fine to program it all on one line. The first step in programming this equation is to set up the variables that are needed for the program, which will be defined as follows:

```
PROGRAM PLC_PRG
VAR
     distance : REAL;
     x1 : REAL;
     x2 : REAL;
     y1 : REAL;
     y2 : REAL;
END_VAR
```

The workhorse of the program – that is, the logic – can be defined with the following code:

```
x2 := 4;
x1 := 2;
y2 := 6;
y1 := 3;
distance := SQRT(EXPT((x2-x1),2) + EXPT((y2-y1), 2));
```

This logic will set the values for the x and y variables and then perform the calculations presented previously. If you programmed the equations correctly, you should be met with the output shown in *Figure 9.8*. As we can see, the calculated value is about 3.61. Typically, for equations like these, you will want to verify your output by plugging the numbers into a calculator. To test and more easily verify the equation, you will want to use very simple whole numbers.

Challenge

As a challenge, modify this program so that each step of the equation is on separate lines. Once you've done that, run the program and compare your results to the previous example code. Are they the same? If not, check your code:

Device.Application.PLC_PRG		
Expression	Type	Value
distance	REAL	3.60555124
x1	REAL	2
x2	REAL	4
y1	REAL	3
y2	REAL	6

Figure 9.8 – Distance calculations

There is no sure-fire way to program an equation. To master this skill, you'll just have to practice numerical programming. As such, we're going to move on to our final projects.

Final projects

As stated earlier this chapter, will have two final projects. To start exploring numerical programming, we're going to look at the final challenge from the previous chapter and program the perimeter of a rectangle.

Final project 1 – programming the perimeter of a rectangle

The perimeter of a rectangle can be described with the following equation:

$$Perimeter \ = \ 2l + 2w$$

This will be a relatively simple equation to program. However, before looking at the code, take a moment and try to solve the problem yourself if you haven't already.

Solution

Our first step is to declare our variables. Since this equation is based around three variables, length, width, and prim, to keep this program as flexible as possible, the data types should be of the REAL type and can be declared with the following code:

```
PROGRAM PLC_PRG
VAR
    prim : REAL;
    width : REAL := 5;
    length : REAL := 6;
END_VAR
```

As can be seen, for this program, we're going to set length to 6 and width to 5.

As we have discussed many times throughout this book, the first step to wiring the program is to break everything down into pieces. The first thing we're going to do is tackle the length calculation. For this, all we have to do is use the following snippet:

```
2 * length
```

We can satisfy the width section of the equation with the following snippet:

```
2 * width
```

Combining the two snippets will render the following logic:

```
prim := 2*width + 2*length;
```

When the program is run, you should hopefully see 22 for prim. Once you have this program worked out, you can move on to the next project.

Final project 2 – Pythagorean theorem

The goal of this program is to write a PLC program that can compute the hypotenuse of a right triangle. Depending on what you're working on and what you're doing, this is a very common program to write. As such, it's a good idea to get some experience writing it now. If you're rusty on geometry, the length of a hypotenuse can be described with the following equation:

$$a^2 + b^2 = c^2$$

In this equation, a and b are the length of the sides and c is the length of the hypotenuse. As with all the other projects, take a pause and try to solve this program before moving forward.

Solution

This equation is dependent on three variables, a, b, and c, all of which make up the hypotenuse. The a and b variables can either be `INT` or `REAL`. However, for this real-world scenario, it would be best to use the `REAL` data type so that they can accept real-world values that will more than likely be floating points. In terms of the hypotenuse, that value should always be `REAL` since it will likely be a floating point due to the `sqrt` function. As such, we can declare the following variables:

```
PROGRAM PLC_PRG
VAR
        hyp     : REAL;
        a       : REAL;
        b       : REAL;
END_VAR
```

Now, the logic for this program can be as follows:

```
a := 5;
b := 6;
hyp := SQRT(EXPT(a,2) + EXPT(b,2));
```

In this case, we are setting the a and b sides of the triangle to 5 and 6, respectively. You can alter these values to any values you want. However, if you keep these numbers, you will get the following output:

Expression	Type	Value
hyp	REAL	7.81025
a	REAL	5
b	REAL	6

Figure 9.9 – Hypotenuse length

To experiment, try changing the structure of the logic and see if you can write the program differently.

Summary

This chapter has been a crash course in the basic mathematics of PLC programming. We have explored the basics of adding, subtracting, multiplication, and division, trigonometric functions, order of operations, and more. Knowing how to program equations into a PLC is a pivotal skill for any automation engineer. At the end of the day, if a PLC programmer cannot program mathematical equations, they can't function as an automation engineer!

One of the biggest takeaways from this chapter is the use of built-in functions. Functions are very important in the realm of PLC programming and many more functions are supported that are non-mathematical. As such, the next chapter will be dedicated to exploring some of the fundamental functions that PLC programmers will often encounter.

Questions

Answer the following questions to test your knowledge of this chapter:

1. What does the floor function do?
2. What is the ABS of -3?
3. What is the ATAN function?
4. How would you write a program that can calculate a quadratic equation?
5. How would you write a program that can calculate 4 to the power of 3?
6. What is the order of operations for a program?
7. What is the assignment operator?

Further reading

IEC 61131-3 Mathematical Functions: https://www.fernhillsoftware.com/help/iec-61131/common-elements/functions-mathematical.html.

10

Unleashing Built-In Function Blocks

As a PLC programmer, whether you're a professional, student, or hobbyist, you've almost certainly run into situations where you needed to copy and paste a series of instructions in multiple places. In programming, having code in more than one location is a terrible practice that should only be exercised in extenuating circumstances. This can lead to a catch-22; on one hand, you may legitimately need the functionality, but on the other, it's a sloppy practice to have the same functionality in more than one location. In that case, what should one do?

A core tenet of any modern programming system is a concept known as code reusability. To accomplish this, entities known as functions and function blocks are employed. Functions and function blocks are complex topics that are core to a quality program architecture. All modern programming languages support some type of "built-in" functionality that is packaged in either a function or function block. This built-in functionality is typically part of what is called a standard library and is often included in a project automatically, while other functionality can be imported manually. A PLC programmer can also create their own custom function blocks.

For this book, we're going to explore built-in function blocks, that is, function blocks that are already included with the CODESYS software. Manually importing function blocks or creating custom function blocks is well beyond the scope of this book; however, knowing how to use the basic functionality that is included in your environment is a must for any PLC programmer. Luckily, if you read the last chapter or you've ever used something like a counter, timer, or sequencer in Ladder Logic before, you're already light years ahead of the game. However, chances are if you're reading this book, you're probably not too sure of how they work under the hood nor how to express those functions in Structured Text.

To explore built-in functions, we're going to explore the following:

- What prebuilt function blocks are
- Internal workings of function blocks

- Rising and falling edges
- Common prebuilt function blocks

To round out the chapter, we're going to utilize a few of the functions and build an industrial washing machine.

Technical requirements

For this chapter, all that will be needed is a working copy of CODESYS installed on your machine. As with all the other chapters in the book, the source code for the projects can be downloaded at the following link:

`https://github.com/PacktPublishing/PLCs-for-Beginners`

It is highly recommended that you pull down the code and try to modify it to gain an in-depth explanation of how the functions work.

What are prebuilt function blocks?

Some operations are so common in PLC programming or programming in general that the developers of the programming tool will often build them into the system. What can be considered a function will vary by name from device manufacturer to device manufacturer; however, for this book, we're simply going to refer to this prebuilt code as functions. Thus far, we have seen prebuilt functions in the form of mathematical operations such as the trig functions, and the more complex math functions, such as the ABS operation and so on. Prebuilt functions extend well beyond mathematical applications and can be found for a variety of tasks. For those who are familiar with Ladder programming, the counter function and the timer function may ring a bell.

There is a difference between a function and a function block in IEC 61131-3. A function is akin to the mathematics functions that were explored in the last chapter. On the other hand, a function block is akin to what is called a class in a modern language like C++, C#, Java, or Python. Not every programming system will support function blocks as they are presented here. Function blocks are an object-oriented programming concept, and that paradigm is novel to the PLC programming world. However, all PLC programming environments will have something akin to functions, which will follow the same general rules as the ones explored here. With that, in terms of usage, what are functions and function blocks and how are they used in a PLC program?

Functions, function blocks, and keywords

On the surface, a function block may look and behave a lot like a keyword. For the inexperienced, this is an acceptable misconception. However, there are key differences between functions, function blocks, and keywords. The following should clarify the difference between the three entities:

- **Keywords**: As explored in *Chapter 3*, keywords are special commands built into the PLC programming language. Depending on the programming system that you're using, the word keyword may be substituted with a synonym, but the core concept is the same. In short, a keyword is simply a special word in a PLC programming language that will perform a specific task such as declaring a variable, comparing two values, creating a function block, or doing any other number of tasks. To invoke a keyword, all one has to do is type it in with the appropriate surrounding syntax when necessary.

- **Function blocks**: A function block is a digital blueprint. A function block offers condensed functionality to perform a task such as creating a timer or a counter unit. Function blocks will offer the ability to cut down on redundant code and allow code portability across compatible PLC code bases. Function blocks have many unique features that neither a keyword nor a function has, such as the ability to share code between different blocks, hiding data in blocks, and so on. To invoke some functionality of a block, all one has to do is create a variable of the function block type name. Once the variable has been created, that variable can be used to access functions and data that is visible in the function block.

- **Functions**: A function is not to be confused with a function block. A function is simply a piece of callable code that can be invoked to do a minor task at any given time. Unlike a function block, a function is not a digital blueprint. Where a function block serves as a larger functional unit, a function is simply a small snippet of callable code that does something such as calculating a special equation. A function can exist independently or inside of a function block. When a function resides in a function block, it is referred to as a method. Though there are some differences in capabilities between a function and a method, they are essentially the same, especially for this book. A function can be invoked simply by calling its name and providing any of the necessary arguments (inputs) that the function requires. If the function is embedded in a function block, the variable that references the function block is required to access the function.

At face value, invoking any of the three entities is as simple as invoking the name. However, that is where the similarities end. A keyword is baked into the programming language and new keywords can't be added, at least not easily. Function blocks and functions can and will usually be added to give a PLC program functionality without the need to add redundant code.

> **Note**
>
> A function is not the same as a function block. A function is a callable block of code that can take multiple inputs if necessary. A function block is a data structure using what are known as classes in a modern object-oriented programming language such as C++, C#, or Java. Many PLC programming environments will support functions, especially mathematical functions, but not all will support function blocks. Some systems may also use different names for both functions and function blocks, but the core data structure will still usually be the same.

A PLC programmer cannot and should not solely rely on keywords. Ironically, some older PLC programmers will avoid the use of functions and function blocks as they view them with suspicion. On the contrary, functions and function blocks are of vital importance in the modern PLC landscape. With the added complexities of the modern automation landscape, code will need to be condensed and reused as much as possible, and the first place to do that is by utilizing prebuilt functionality.

To some, there still may be some confusion as to what a function block is. Therefore, to get a better idea of what a function block is and what prebuilt function blocks offer, let's explore the guts of a function block.

Function blocks under the hood

Thus far, we've explored function blocks a little, mostly how they work at a high level. However, only knowing how a function block works in practice will only get you so far. To fully understand what a function block is, we need to explore function blocks at a conceptual level.

What is a function block?

Okay, so we have established that a function block is analogous to a class in C++ or Java, and that a function block is a digital blueprint, but what exactly does that mean? Well, in short, a function block is a thing and contains all the necessary support logic for that thing to function. For example, consider a car. For a car to work, it needs things such as an engine, wheels, and a chassis. It will also need certain data such as the number of miles the car gets per gallon of gas. If you think about it in terms of manufacturing, you don't want to have to draw a blueprint for each car that comes off an assembly line. Instead, you want to build a series of cars based on the same blueprint. A function block is similar to this. You have multiple references, such as multiple timers or counters that utilize the same code. Essentially, you can have `timer1` and `timer2`, which are both derived from `TON`.

A function block works the same way. A function block will contain all the necessary logic and data for a specific functionality. For example, a timer will have all the necessary variables and functions to trigger a bit after a given amount of time. In other words, a function block is a series of contained functions and variables that work in unison to accomplish a goal.

When correctly architected, a function block will do the following:

- Reduce redundant code for a given project
- Allow functionality to be ported across compatible projects
- Provide all the necessary functionality and data to perform a certain task

Some of these prebuilt functions may seem a bit odd to use at first. This is mostly because they rely on what is known as a rising or falling edge. For entry-level PLC programmers and automation engineers, the whole concept of rising and falling edges can be a bit confusing, so we're going to dedicate the next section to exploring rising and falling edges.

Rising and falling edges

A lot of functionality will depend on what is known as a rising or falling edge. In the most lay sense, a rising or falling edge can be thought of as some type of user action such as pressing or releasing a button. This may seem trivial, but this is very important. For example, consider a counter; it needs to know when to increase. For some applications, the incrementation may need to occur during a button release, and for others, it will need to occur during a button press.

Conceptually, a rising and falling edge can be seen graphically in *Figure 10.1*:

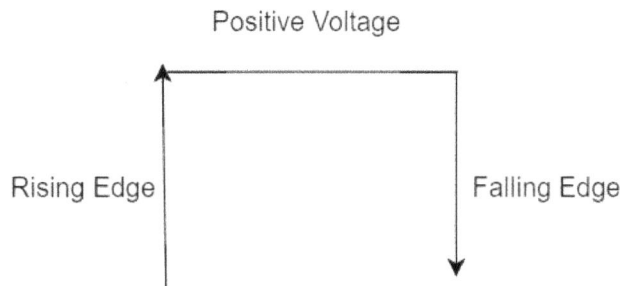

Positive Voltage

Rising Edge   Falling Edge

Figure 10.1 – Rising and falling edge

Essentially, a rising edge can be thought of as a rise to the peak voltage such as pressing a button to create a closed circuit. On the other hand, a falling edge is a discharge from a positive voltage to 0v when a button is released, and the circuit is opened.

> **Note**
> A rising edge is the result of something such as a switch closing and energizing a circuit, while a falling edge is akin to a circuit opening and a circuit de-energizing.

A logical question is why is this relevant in a discussion about functions and function blocks? As stated before, many functions will work whether a rising or falling edge is detected. Again, this stems back to certain functions needing to know whether to increment when a button is pressed or a button is released. Understanding what rising and falling edges are is vital to the next steps in the exploration of functions and function blocks.

The role of rising and falling edges will vary depending on the function or function block. So, what are some function blocks that an IEC 61131-3 PLC programmer uses in their day-to-day tasks? The next section is dedicated to answering that question, as in the next section, we're going to explore some of the most common function blocks that a PLC programmer will interact with.

Common PLC function blocks

As with anything else in programming, there are certain features that you'll use more than others. Depending on what job you're working on, you're likely to use either a counter or a timer. Let's explore a counter function block.

Counter function blocks

As the name suggests, a counter block counts. These function blocks are often used to measure the number of parts that pass through a given point, the number of times a button has been pressed, the number of times a machine has been started, or any other counting operation. The easiest way to explore this function block is to see it in action. The first thing we're going to do is create the following variables in the code snippet:

```
PROGRAM PLC_PRG
VAR
     counter     : CTU;
     count       : REAL;
     buttonPress : BOOL;
END_VAR
```

In this variable list, the heart and soul is the `counter` variable. As can be seen, this variable is of type CTU. This means the counter variable is a reference to a counter function block that that counts up. There is another type of counter, CTD, that counts down. For this book, we're only going to explore the CTU block as both blocks work in a similar manner. A function block can be thought of as a special data type; however, we're going to leave it at that for now because more complex knowledge of object-oriented programming is needed to fully understand this, which is beyond the scope of this book. The next variable (`count`) will keep track of the number of increments, and the `buttonPress` variable will act as our rising edge.

Once you have the variables set up, you can move on to the core logic of the program, which should match the following:

```
counter(CU := buttonPress);
count := counter.CV;
buttonPress := 0;
```

In this code snippet, `counter` is what is responsible for creating the timer. In terms of Ladder Logic, this line is the same as dropping in a counter instruction like in *Figure 10.2*, only with more of the variables filled in:

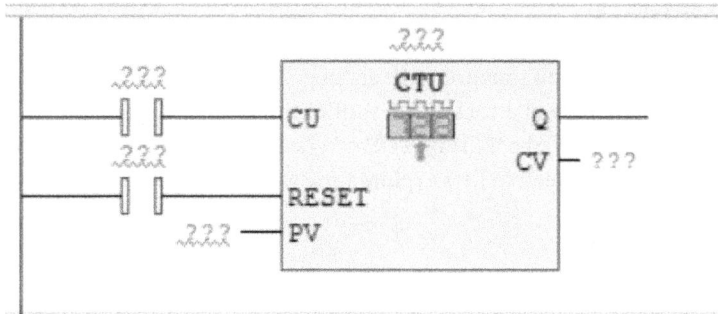

Figure 10.2 – Empty CTU function block

Now, notice in *Figure 10.2* that there are multiple inputs and outputs, such as the PV, CU, RESET, and so on. The text-based version of the CTU block will also have these, but for the text-based version they are known as arguments, and they can be viewed in the core logic code snippet. As can be seen in the code snippet, each one of these inputs can be set in parentheses with the following syntax:

```
Counter_var(input_name := variable, …);
```

Like with the Ladder version of the function block, you will only need to set the fields that are relevant to your project. For this example, we are only going to set a buttonPress variable to the CU field. In other words, in this case, every time there is a rising edge for the buttonPress variable, the counter will increase by exactly one.

In terms of the outputs, we need to worry about the CV field. This field holds the current count. As in the code snippet, we created a variable called count, which will be assigned the current count from the counter variable. In all, when the code is run, you should see something like the following:

Figure 10.3 – Counter example

To increment the counter, simply write a TRUE to the buttonPress variable. Notice that each time you write TRUE to buttonPress, the value will increment by one.

A PLC programming professional will often need to use more function blocks than just the CTU block. Another common function block that a PLC programmer will use is a timer block.

Timer function blocks

Often, things have to be timed in automation. Mixers may need to run for a pre-defined amount of time, a batch mixer may have to wait a certain amount of time before it can start, and so on. To do this, a prebuilt function block called a TON or TOF can be used. However, before we dive into the actual timer function blocks, we need to first explore time variables.

Exploring time variables

For a timer to work, we need to tell it how long to wait before it turns on or off. This is unlike most other programming languages where you can simply pass a number to a special function that will pause the program. However, in the IEC 61131-3 standard, there is a special syntax and data type that are used specifically for timing intervals.

First and foremost, the IEC 61131-3 standard has a special data type called time that is designed to support temporal units. To declare a time variable, you will use the TIME data type. For example, if one wanted to create a variable called delay that is meant to be used for a timer, you would use the following syntax:

```
delay : TIME;
```

Now, actually assigning a time value to this variable can be a bit tricky, as it will have a unique format. Essentially, the temporal value will match the following format:

```
T#<Time><time_unit><optional_time><optional_time_units>
```

If you needed 500 ms, you would use the following:

```
delay := T#500MS
```

Now, if you needed 2 seconds and 500 ms, you would use the following syntax:

```
delay := T#2S500MS
```

Once you understand this basic syntax, you will be able to move on to utilizing these temporal units for timers.

Exploring the TON function block

The first timer function block that we're going to explore is the TON block. The TON or Timer On will turn a bit on after a given amount of time. These function blocks are good for delaying something such as the start of something like an industrial oven, mixer, or the like. To demonstrate this, let's set up the following variables:

```
PROGRAM PLC_PRG
VAR
    timer : TON;
```

```
      delay : TIME;
      in    : BOOL;
      out   : BOOL;
 END_VAR
```

In this case, the variable timer is a reference to the TON function block, the delay variable will be used for timer wait time, the in variable will serve as the rising edge, and finally, out will be the bit that we are turning on when the timer turns on.

Once you have those variables in place, you can set up the core logic of the program with the following:

```
delay := T#2S500MS;
timer(IN := in, PT := delay);
out := timer.Q;
```

The first line will set the delay (time to wait until the timer is on) to 2 seconds and 500 milliseconds. Line two is the setup for the TON function block. Finally, line three is simply the state of the time, that is whether or not the bit has turned on or not yet. In the case of a time, the Q output is the state of the timer. When the timer is off, Q will be false, and when the timer fires, Q will be TRUE. So, essentially, Q is the state of the timer.

To test this program, simply run the program and write the in variable as TRUE. Once you have waited the allocated amount of time, that is, approximately 2.5 seconds, the out variable will be set to TRUE and the output should match *Figure 10.4*:

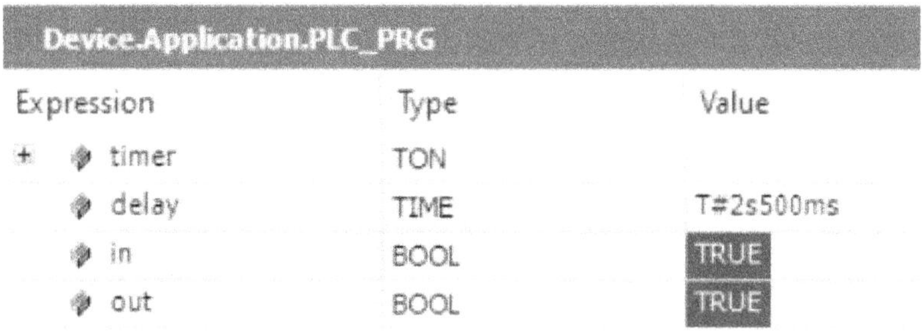

Device.Application.PLC_PRG		
Expression	Type	Value
+ ♦ timer	TON	
♦ delay	TIME	T#2s500ms
♦ in	BOOL	TRUE
♦ out	BOOL	TRUE

Figure 10.4 – TON in on state

If you're not familiar with how these bits change, the transition can be easy to miss, and as a sanity check, you may want to run the timer a few times to really observe the change. To reset the timer, all you need to do is set the in variable to FALSE and then set the in variable back to TRUE. You may also want to change the delay variable to a larger or smaller value for exploratory purposes. Now that we've explored the TON timer, we need to move on to its sister function block and explore the TOF function block.

Exploring the TOF function block

The TOF function is very similar to the TON function block, but it works in reverse. Instead of the output bit being off by default as with the TON function block, the TOF will remain on or TRUE (when the input bit activates the timer and then goes to false) until the allocated time unit has passed, and then it will toggle to FALSE. A TOF is often used for running processes for a given period. For example, if a mixer needs to run for 30 minutes, it is common to use a TOF with a 30-minute time allocation. There is another gotcha to a TOF; unlike a TON, which requires the in bit to be TRUE, a TOF will only activate when the in variable is set to FALSE.

We're going to use the same code as we did before, with one minor modification to the variables portion. In short, your variable section should look like the following:

```
PROGRAM PLC_PRG
VAR
     timer : TOF;
     delay : TIME;
     in    : BOOL;
     out   : BOOL;
END_VAR
```

For the same core logic, nothing will change and we're going to use the following:

```
delay := T#2S500MS;
timer(IN := in, PT := delay);
out := timer.Q;
```

When the program is running, set the in variable to TRUE. Then, set the in variable to FALSE and observe the behavior of the out variable. After about 2.5 seconds, the variable should toggle itself off.

There are many more prebuilt function blocks that can be utilized in a PLC program. The CTU, TON, and TOF function blocks are just three very common examples of heavily used function blocks. CODESYS comes built with a few more and supports a very large ecosystem of libraries and third-party modules that can be utilized in your project. Examples for function blocks that can be used include the following:

- Cloud computing
- Machine learning
- Reading/writing CSV files
- Motor drives
- Complex mathematics

There exists a function block for pretty much anything you could possibly need. For now, we're going to move on to our final project and create an industrial washing machine.

Challenge

More often than not, you will run into a function block that you don't know how to use. This means that you will have to research how to use it and use it effectively. A good example of this is the CTD function block that was mentioned before. Before you move on to the final project challenge, research the CTD function block and try to use it in a program that can count down from 100 to 0.

Final project

Industrial washing machines are common applications for timers. Typically, an operator will press a button and the washing machine will run for a given amount of time; then, a spin cycle will start. Essentially, this project will be an exercise in cascading timers. When one timer fires, it will trigger the start of another timer. Therefore, to begin this project, we will need the following variables.

Variables

To start off, let's declare the following variables:

```
PROGRAM PLC_PRG
VAR
        wash        : TON; //washing machine timer
        spin        : TON; //spin timer
        start       : BOOL; //simulated start button
        washOn      : BOOL; //washing state
        spinOn      : BOOL; //spin state
        delay: TIME; //delay
END_VAR
```

Take a second to study what the variables do. Essentially, start will trigger the whole washing process by triggering the timers. When you have the variables in place, implement the following for the core logic:

```
delay := T#1S;
wash(IN := start, PT := delay);
washOn := wash.Q;
spin(IN := wash.Q, PT := delay);
spinOn := spin.Q;
```

As can be seen in the logic, we will run the wash cycle for 1 second then the spin cycle will start after 1 second. To test the program, run the program and pay attention to the washOn and spinOn variables.

When the program is running, you should see something like the following after about 2 seconds of changing the start variable to TRUE:

Device.Application.PLC_PRG		
Expression	Type	Value
+ ◆ wash	TON	
+ ◆ spin	TON	
◆ start	BOOL	TRUE
◆ washOn	BOOL	TRUE
◆ spinOn	BOOL	TRUE
◆ delay	TIME	T#1s

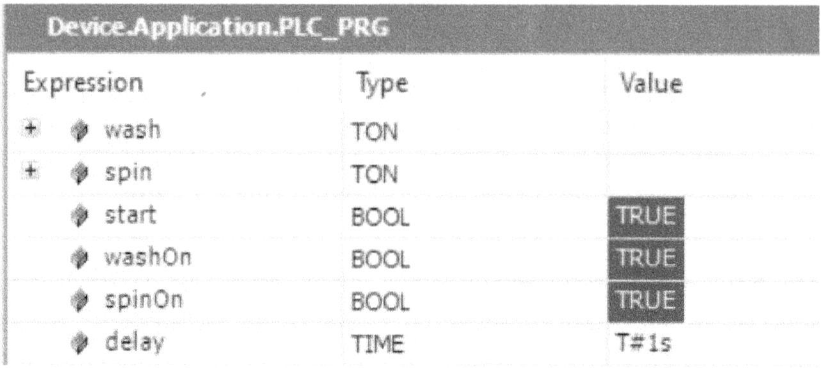

Figure 10.5 – Washing machine

To restart the process, simply toggle the start variable back to FALSE and set the start variable to TRUE again.

Summary

In this chapter, we explored pre-built function blocks such as the CTU, TOF, and TON blocks. We have also explored the time data type and the syntax to create a time variable. The key takeaway for this chapter should be that there are a lot of prebuilt functionalities that can make life as a PLC programmer very easy. Another key takeaway should be that each function block has different inputs and outputs. This means that learning how to use function blocks will take a bit of research.

Thus far, all of our programs have flowed very linearly. That is, all our programs have flowed from top to bottom with no branches involved. Though we can add a lot of advanced functionality to our programs with function blocks, our programs will always lack any real intelligence due to the straight linear flow. As such, in the next chapter, we're going to explore flow control and how we can control the output of our program based on certain inputs.

Questions

1. What is a function block?
2. Can a function live in a function block?
3. Do all PLC programming systems support function blocks?
4. Name three common function blocks in CODESYS.

5. Do all function blocks have the same inputs and outputs?

6. What is the main difference between a TON and a TOF timer?

7. What is the main difference between a timer and a counter?

Further reading

- CTU CODESYS documentation:

  ```
  https://help.codesys.com/webapp/
  ctu;product=codesys;version=3.5.11.0
  ```

- TOF CODESYS documentation:

  ```
  https://help.codesys.com/webapp/
  tof;product=codesys;version=3.5.11.0
  ```

- TON CODESYS documentation:

  ```
  https://help.codesys.com/webapp/
  ton;product=codesys;version=3.5.11.0
  ```

11
Unlocking the Power of Flow Control

So far in this book, we've only explored programs with a linear flow. That is, all the programs that we explored have started at the top and ended at the bottom. No matter what, this is the natural flow of all software. However, the route we take to get to the bottom will often vary.

Typically, a program will have no one defined path to the bottom. A program will usually branch off into multiple paths to get to the bottom. These paths will have different logic that will be executed, which will alter the program's behavior depending on which path it takes. To conceptualize this, consider the logic that turns on a motor. The type of VDF a customer chooses will often dictate a start sequence. A quality program will either detect the motor brand or allow the operator to input the motor brand, and the program will run the proper logic based on the selection.

Flow control is a very important topic in both traditional and PLC programming. As with many of the other topics that have been covered so far, if you do not understand flow control you will not succeed as a PLC programmer. To remedy this and help you master the concept, we're going to cover the following:

- Exploring what a control statement is
- Exploring why flow control is important
- Exploring logical expressions
- Exploring the IF statement
- Designing a program using flow control
- Exploring the CASE statemen

Finally, to round out the chapter, we are going to design and build a color sorting program that can sort based on color input. This chapter is going to depend heavily on pseudocode and flowcharting; as such, if you do not have a quality grasp on those concepts, you need to review *Chapter 5*.

Technical requirements

For this chapter, you will need a working copy of CODESYS installed on your machine. As with all the other chapters, this chapter will have example code that can be downloaded at the following URL:

`https://github.com/PacktPublishing/PLCs-for-Beginners`

As with the other chapters, it is highly recommended that you download and experiment with the code to better grasp the material presented in this chapter.

Exploring what flow control is

Similar to how your behavior varies based on certain conditions, so too will your program. Flow control is the mechanism by which your program will choose what route to take to get to the bottom. Consider the famous beer example that we explored earlier in the book:

```
Input age
If age >= 21 then
     Let them drink heavily
Else if age < 21 then
     Throw a shoe at them and call the cops
```

In this case, there are two paths that the program can take. One path will allow a person to drink heavily, while the other path will throw a shoe at the perpetrator and have the operator call the cops. This is the heart of flow control. Essentially, there are multiple paths, and the PLC will choose the correct path to take based on the given inputs: in the case of the beer program, the person's age.

In terms of decision based flow control, there are two special keywords that the *IEC 61131-3* standard supports. The two keywords are as follows:

- `IF`
- `CASE`

There are a couple of other derivatives of the `IF` statement that are also supported, but those will be explored in the next chapters. The `IF` and `CASE` statements will ultimately do the same thing, but they work differently. We're going to explore both keywords later on in the chapter, but for now, we're going to explore why we need flow control for a program.

Exploring why flow control is important

As stated before, a program will rarely take a straight path from top to bottom. The necessary logic that needs to run will often vary based on a given input. Think about the motor example at the beginning of the chapter. For different VDFs to operate properly, they need certain logic such as special startup commands, startup sequences, and so on. Flow control is vital for applications such as this. The motor type can be considered the input, and the specialized logic is the varying output.

A more conceptual application for flow control is the aforementioned beer-buying example. In short, that program determines if a person can buy beer or not. A user is inputting an age, and the program determines if a person can drink. This program may seem trivial, but applications such as this are very common in the PLC realm. It is very common to have to take an input value such as a motor's voltage, current, or temperature and adjust the VDF's parameters based on that input.

Essentially, flow control helps with the following:

- Adjusting operational parameters based on a given input
- Running or ignoring logic blocks based on inputs or internal conditions
- Modifying the general behavior of a program based on certain conditions

With that, what are the basics of flow control? How does one program these conditions? The first step in doing that is understanding logical expressions.

Exploring logical expressions

For a control statement to work, it will need some type of logical expression to evaluate to True for the code in the statement to be executed. A logical expression can be thought of as something akin to comparing two numbers to see if they are equal, if one number is less than the other, or if one number is greater than the other. The *IEC 61131-3* standard uses the following for logical comparisons:

- **Equal**: The equals operator compares two values and will evaluate to True if the two numbers are equal. In *IEC 61131-3*, the equals operator is denoted with the = symbol.

- **Not equal**: The not equals operator is the inverse of the equals operator. In short, when the two values in the expression are not the same, the expression will return True. Not equal is denoted by the < > symbol in *IEC 61131-3*.

- **Less than**: The less-than operator will check if one number is less than the other. If the number on the left is less than the number on the right of the symbol, the expression will evaluate to True. Less than is expressed with the < operator. An example expression would be 2 < 3. In this case, the expression would return True.

- **Greater than**: The greater-than operator works the exact same way as the less-than operator and is denoted with the > symbol. Unlike the less-than operator, if the number on the left is greater than the number on the right, the expression will return True.

- **Less than or equal to**: This expression will evaluate to `True` in two circumstances. The expression will evaluate to `True` when the value on the left is less than the one on the right or is equal to that value. For *IEC 61131-3*, the symbol to denote this is `<=`.

- **Greater than or equal to**: This expression is very similar to the less-than or equal to expression, except it will evaluate to `True` when either the value on the left is greater than the value on the right or equal to it. The symbol to denote this is `>=`.

These logical expressions must be memorized at all costs to be successful as a PLC programmer because they are not only used in decision statements but loops as well.

These expressions will all return a Boolean value. These expressions will either return `True` or `False`. Therefore, you assign the output to a variable or simply use them in a loop or control statement. To see these expressions in action, we're going to explore the almighty `IF` statement.

Exploring the IF statement

The most rudimentary control statement that all PLC developers will need to master is the `IF` statement. As the name suggests, this command will run a block of code if, and only if, a certain logical expression evaluates to `True`. Much as with logical expressions, the `IF` statement must be mastered, especially for PLC programming.

IF statement syntax

The *IEC 61131-3* syntax for the `IF` statement is somewhat like the `VAR` declaration block. The basic structure for the `IF` statement is as follows:

```
IF <expression> THEN
    //code to run
END_IF
```

For this structure, the expression is a logical expression such as determining if two values are equal. In between `IF` and `END_IF` is what's called the body. The body is the code that will run when the expression is `True`. Technically, you can have as much code in the body as you want, but it is advisable to keep the body as short as possible. To really understand the `IF` statement, we need to implement a working example to explore.

Checking if two values are the same!

The easiest and arguably most logical way to explore the `IF` statement is to see if two numbers are equal. For many, this is their first experience with the code structure, and, in many ways, it can be thought of as the *Hello World* of control statements. To explore the equals operator, we're going to write a simple program that can determine if one variable is 3 or not. For this program, we're going to declare two variables. We're going to declare a simple integer variable that we'll call x that we're going to initialize to 3 and a bool variable that we'll call state and initialize it to False. When implemented, the variable section of your program should look like the following snippet:

```
PROGRAM PLC_PRG
VAR
    x : INT := 3;
    state : BOOL := FALSE;
END_VAR
```

The logic for this program will match the following:

```
IF x = 3 THEN
    state := TRUE;
END_IF
```

The way this program will work is simple. When x is 3, the state variable will toggle from False to True. When the program is run, you should get the output that can be seen in *Figure 11.1*:

Figure 11.1 – IF statement execution

To demonstrate how the IF statement would not execute, modify the logic to match the following:

```
IF x = 5 THEN
    state := TRUE;
END_IF
```

In this example, all we did was change the logic expression from x = 3 to x = 5. If nothing else is changed, the expression will evaluate to False since x is still set to 3. As such, the state variable should not change. When the code is run, you should be met with the output seen in *Figure 11.2*:

Figure 11.2 – IF statement False condition

As can be seen in *Figure 11.2*, the state variable did not change from `False` because the code was never run due to the expression in the `IF` statement evaluating to `False`.

In many circumstances, we want to determine if two values are not equal. In other words, there are situations when we want the state to be `True` when x is not equal to 5. Basically, we want the inverse of what we got for *Figure 11.2*. To do that, we would use the not equals operator, which we're going to explore next.

Exploring the not equals operator

As stated at the end of the last section, it is very common to need to run a block of code when two values are not the same. This means that there will be times when we want the `IF` statement to evaluate to `True` when the two numbers are not the same. This may seem oxymoronic, but it is a very common application. For example, if you have a machine with a proximity switch that will produce a `1` or `High value` when the sensor is tripped, you will generally want to shut down the machine. Otherwise, you'll want the machine on. You can use something like the following to accomplish the task:

```
Sensor_Input = input from sensor
If (Sensor_Input <> 1) then:
    Turn on machine
Else if (Sensor_Input == 1) then:
    Turn off machine
```

Now that we have an idea of what the pseudocode looks like, let's move on to a working example.

Real code example

To demonstrate the not equals operator, we're going to use the same variables from the equals example but modify the core logic to match the following:

```
IF x <> 5 THEN
    state := TRUE;
END_IF
```

The x variable should be initialized to the integer 3 as it was before. So, when the program is executed, we should be met with the output seen in *Figure 11.3*:

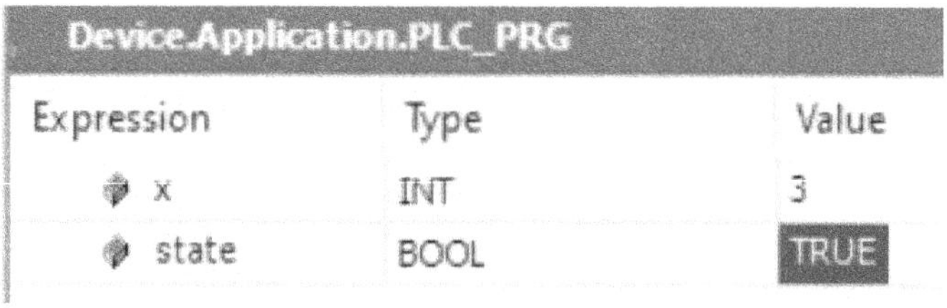

Device.Application.PLC_PRG		
Expression	Type	Value
⬦ x	INT	3
⬦ state	BOOL	TRUE

Figure 11.3 – Not equals output

The output from this example is the inverse of what we saw before. In the previous example, the state variable was toggled to `True` when the values were equal and stayed `False` when they were not. In this example, the state variable was toggled to `True` when the two numbers were not equal.

A program can have as many `IF` statements as is necessary to complete the job. However, a bunch of `IF` statements can make a code base messy and unmaintainable. Therefore, there will be a balancing act between having enough `IF` statements to implement the code while keeping your code base clean and maintainable.

Thus far, we have only explored the equals and not equals logical operators. However, there are still a few more operators that were mentioned in the *Exploring logical expressions* section. Take a moment to see if you can solve the following challenge.

Challenge

In terms of syntax, the logical operator will affect nothing. In other words, the same general structure of the `IF` statement will not change. The only thing that will change is the operator you will use. So, for a challenge, write a program with the following requirements:

- The program should have two `BOOL` variables called `lessThan` and `greaterThan`. There should also be one `INT` variable called `input`.
- If the input is less than 5, set `lessThan` to `True` and all others to `false`.
- If the input is greater than 5, set `greaterThan` to `True` and the rest to `false`.

There is no right or wrong answer to this challenge. The only thing that matters is whether the program works or not. Do your best to implement this program; take as much time as you need. Once you complete this challenge, feel free to move on to designing a program with control statements.

Designing control statements

Flow control can be somewhat confusing at first. It can be difficult for inexperienced programmers to visualize the flow of a program. This is where flowcharting can help new programmers visualize the flow of the program. Flowcharting may appear to be a pointless design technique. However, it can greatly help alleviate confusion and make the flow of the code more understandable where multiple flow paths exist. To better under this concept, let's design and implement a simple bank account overdraft program.

Overdraft program

Most bank accounts will throw some type of warning message when a user withdraws too much money from their bank account. In real-world practice, a lot goes into the software that governs bank accounts; nonetheless, a simple `IF` statement can easily simulate the same behavior. The first phase in implementing this program is designing the flow of the program.

Program design

For this program, we can use a design like the one in *Figure 11.4*:

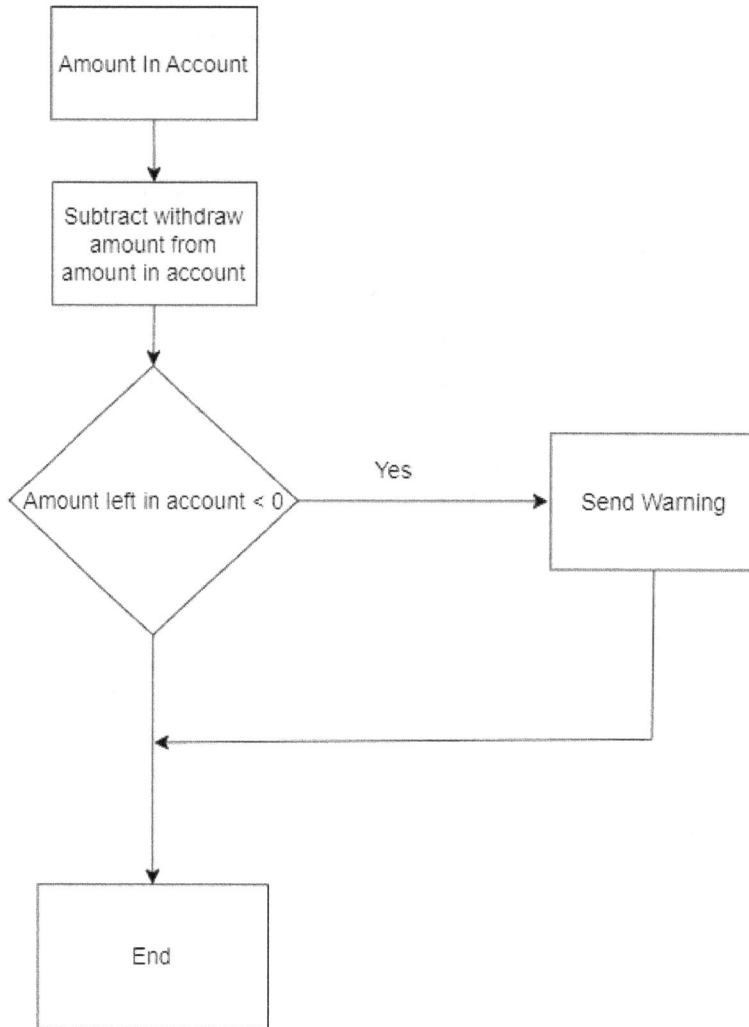

Figure 11.4 – Program design

In this design, we will start off with a given amount in the account, then we will withdraw another amount. If the amount of money is less than 0 (overdrawn), the program will send a warning and then end. If the amount is not less than 0, the program will simply end. Now, since this is a PLC program, the software will not end but loop; this is important to remember since we are writing a program that will live and run on a PLC.

Code implementation

Now that we have a design to work from, we can start implementing the code. As usual, the first thing we need to do is declare the variables. For this example, we're going to use the following:

```
PROGRAM PLC_PRG
VAR
     initalAmount        : INT;
     withdrawAmount      : INT;
currentAmount           : INT;
     message             : WSTRING;
END_VAR
```

For this example, the `initalAmount` variable will hold the amount of money that we're starting off with, the `withdrawAmount` variable will represent how much we're going to take out of the account, and the `currentAmount` variable will represent the amount left in the account after a withdrawal. The `message` variable is there to alert the user if the account has been overdrawn or not.

Now, before you proceed, try to write the program from the flowchart design; we're starting off with $200 in the bank account, and we are withdrawing $100. Remember there are no right or wrong answers, as long as the program works. After you take a stab at the PLC code, compare your answer to the following:

```
initalAmount    := 200;
withdrawAmount := 100;
currentAmount   := initalAmount - withdrawAmount;

IF currentAmount < 0 THEN
     message := "you are overdrawn";
END_IF
```

When this code is run, you should get the values shown in *Figure 11.5*:

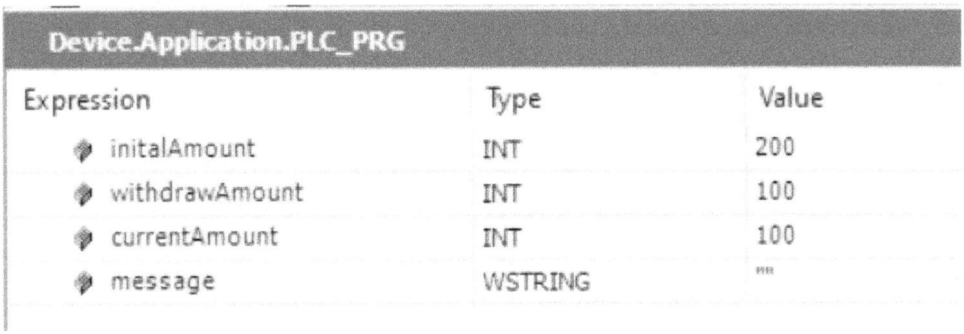

Device.Application.PLC_PRG		
Expression	Type	Value
initalAmount	INT	200
withdrawAmount	INT	100
currentAmount	INT	100
message	WSTRING	

Figure 11.5 – No overdraft output

In this case, there was no overdraft. This is because we had $200 in the bank, and we only took out $100. Now, to test the overdraft warning, modify the code to match the following:

```
initalAmount    := 200;
withdrawAmount := 250;
currentAmount  := initalAmount - withdrawAmount;

IF currentAmount < 0 THEN
    message := "you are overdrawn";
END_IF
```

In this case, all we did was change 100 to 250 on the second line. Though a small change, this change will have a profound impact on the behavior of the program. After you modify the code, run it, and you should be met with the same outputs that are presented in *Figure 11.6*:

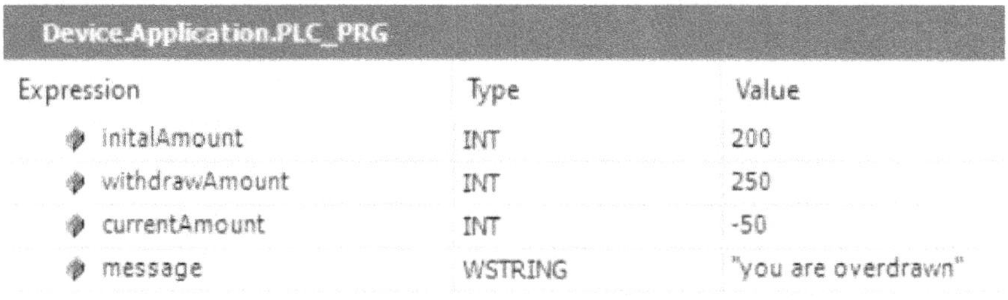

Device.Application.PLC_PRG		
Expression	Type	Value
initalAmount	INT	200
withdrawAmount	INT	250
currentAmount	INT	-50
message	WSTRING	"you are overdrawn"

Figure 11.6 – Overdraft notice

This means that our design worked, and we successfully implemented a program from a flowchart. Now, this may have seemed too trivial for a flowchart, especially if you've worked with IF statements in the past. However, for large complex control structures such as that of a state machine, learning how to implement a design from a flowchart is quite beneficial as it'll help you keep track of the larger picture.

By this point, you should have a pretty good feel for how IF statements work. As stated before, IFs are not the only type of control statement. In fact, IF statements can be awkward and even detrimental for certain applications such as state machines, which will often use a numerical code to determine which block of code to run. This mostly stems from a much larger code overhead required for the IF statement to properly run. In cases such as these, there is a much better alternative called the CASE statement that we're going to explore in the next section.

Exploring the CASE statement

If you've ever programmed in a language such as C/C++, Java, C#, or any other modern programming language, you might be familiar with a SWITCH statement. The *IEC 61131-3* standard doesn't support a SWITCH statement per se, but it does support a CASE statement, which is, for all purposes, the same. The CASE statement is like the IF statement, except it requires less overhead in terms of code.

However, there is a drawback to the CASE statement in that it can only check for equality. This makes the CASE statement ideal for applications that require comparison of data such as status codes but less than ideal for applications that check for inequality.

State machines and CASE statements

A common application for the CASE statement is what is known as a state machine. State machines are an advanced concept that are only going to be touched on here. There is a lot of theory that needs to be understood to fully understand state machines. Therefore, we're only going to explore them at a high level in this book.

A state machine can be thought of as a code block whose output is determined by its input. Consider the following pseudocode:

```
Input state //numerical value 0 = off, 1 = on, 2 = standby
CASE state
     Case state = 0:
          Turn machine off
     Case state = 1:
          Turn machine on
     Case state = 2:
          Put machine in standby mode
```

As can be seen in the example, when the state equals a certain value, a certain block of code will be executed. This is a very simple state machine that uses a CASE statement.

All things considered, the example pseudocode is easy to follow. Unlike most other things in programming, the actual implementation of a working CASE statement is much simpler. The basic syntax for the CASE statement is the following:

```
CASE <variable> OF
     <val_1>:
          Code
     <val_2>:
          Code
     <val_N>:
          Code
END_CASE
```

In the next section, we're going to put this syntax to work and turn the pseudocode design into a working state machine!

Implementing the state machine

To begin, we're going to start off with two variables: one called `input` of type `INT` and another called `state` of type `WSTRING`. To do this, match your variable section with the following code:

```
PROGRAM PLC_PRG
VAR
    input : INT;
    state : WSTRING;
END_VAR
```

For this example, the `input` variable will be used to take in a state. For the program, we will use the same number codes that were used in the pseudocode example, while the output string will hold the current machine state, as with the pseudocode example. Once you implement your variables, you can move on to implementing your core logic, which should look like the following:

```
CASE input OF
    0:
        state := "Machine OFF";
    1:
        state := "Machine ON";
    2:
        state := "Machine STANDBY";
END_CASE
```

When you start the program and set `input` to 0, you should be met with the following outputs:

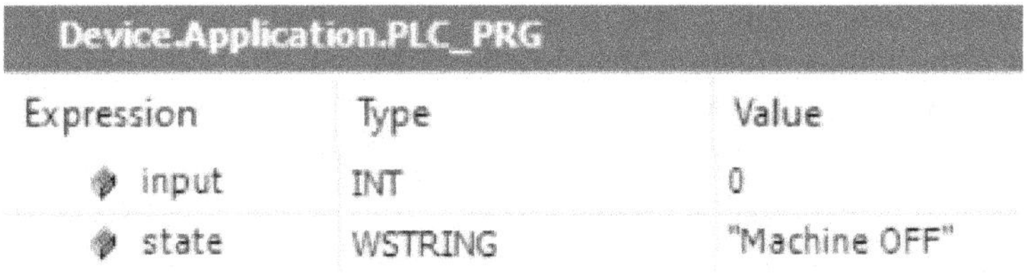

Device.Application.PLC_PRG		
Expression	Type	Value
input	INT	0
state	WSTRING	"Machine OFF"

Figure 11.7 – Machine OFF state

Now, write the value 1 to input, and you should be met with the output seen in *Figure 11.8*:

Device.Application.PLC_PRG		
Expression	Type	Value
◈ input	INT	1
◈ state	WSTRING	"Machine ON"

Figure 11.8 – Machine ON state

Finally, when you write the number 2 to input, you should be met the output shown in *Figure 11.9*:

Device.Application.PLC_PRG		
Expression	Type	Value
◈ input	INT	2
◈ state	WSTRING	"Machine STANDBY"

Figure 11.9 – Machine standby state

Now, this program could have been accomplished with a series of IF statements; however, as can be observed, this is a much cleaner and more concise approach.

Flowcharting and CASE statements

CASE statements are for code blocks with a lot of would-be IF statements. It is meant to keep the code base cleaner and easier to follow. This usually means that there are many more branches for the program to follow. As a result, it is wise to have a visual aid to follow when working with applications such as state machines. As I'm sure you've guessed, this is where flowcharts come in handy. If we were to flowchart the state machine we just implemented, it would look something like the following:

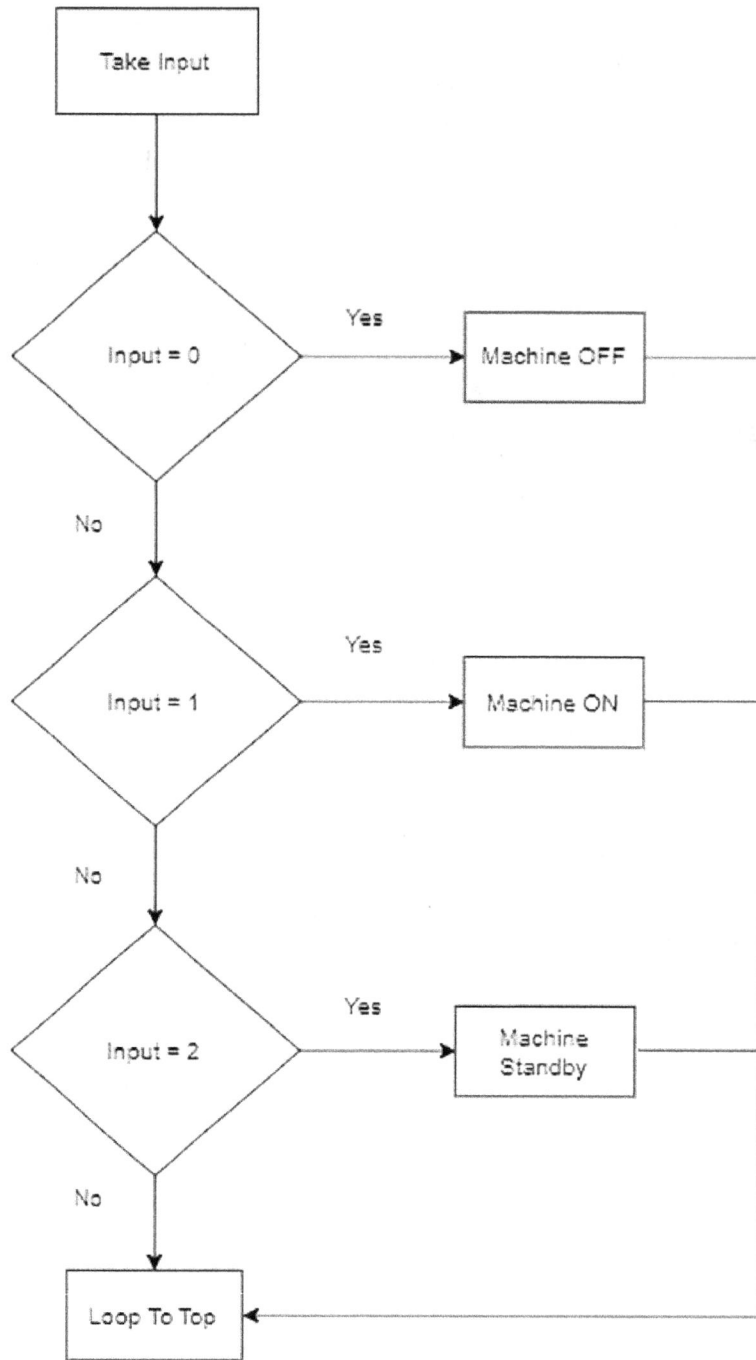

Figure 11.10 – State machine flowchart

As can be seen in *Figure 11.10*, the program will keep looping regardless of the input state. However, when one of the conditionals is returned `True`, it will change its message and then loop again.

> **Note**
>
> Flowcharting is an excellent way to design complex control statements. However, many experienced engineers will either skip the design phase altogether or use pseudocode for the design. Flowcharting usually helps lay people, such as customers, to better understand how an operation will work and is a good idea to use.

A logical question that many students and inexperienced programmers will often ask is, What is a real-world application for the `CASE` statement? In other words, what are some common examples that the `CASE` statement will be used for?

Real-world applications for the CASE statement

First, the example that we explored is a real-world example. Having a machine's state be dictated by an input is a common `CASE` statement application. It is very common for engineers to set up a series of codes that when fed into a `CASE` statement will perform certain power-on or power-off functions for a machine or machine feature. For example, a programmer may rig up a system that, when fed in the number 1, will start the machine, and then rig up another `CASE` statement that, when fed in the number 2, will start a saw.

Another common use for a `CASE` statement is with status codes. Typically, VDFs and other devices will send status codes to the PLC via some type of communication protocol. Once the PLC picks up the status code, it is common to use a state machine with a `CASE` statement, such as the one presented here, to run the necessary code to accommodate the status. For example, if an error code is 999, a PLC programmer will usually set up a case for 999 and do something such as turn off that operation or put the machine in standby mode.

By this point, you should have a decent understanding of how flow control works. Let's now move on to our final project and implement a color-sorting device, which is a device that is becoming very common in the automation world!

Final project

Color sorting has become a staple of modern automation. Often, machines will tag parts with different colored stickers that will dictate their fate. For example, a red sticker may mean *reject*, a blue sticker may mean *Line B*, and a green sticker may mean *part accepted*. It'll all depend on the manufacturing process. For our final project, we're going to make a simulated color sorter to help sort boxes based on sticker color.

Requirements

The following are the actions and conditions that the machine will need to take:

- The machine will have an input variable that will dictate which state the machine is in. When the input is 1, the machine will turn on; when the input is 2, the machine will turn off; and when the input is 3, the machine will go into standby mode.

- When the machine is running, it must read the color-code variable that will store the color code of the sticker it is reading.

- If the color code is red, we need to reject the part; if the color is green, we can accept the part; and finally, if the color is yellow, we will send the part back for inspection.

Now that we have a few basic requirements, we will move on to designing the program.

Program design

This program may sound intimidating, but it is quite simple. Most color detection modules work off an RGB system that will return a value that is indicative of the actual color. For our purposes, we're going to assume that 1 means red, 2 means green, and 3 means yellow. In a real-world application, the color codes would be more complex, but for our purposes, we're going to keep things simple and use only a single digit.

The first task that we need to tackle is coming up with a design for the program. You should take a moment to try to draft out a pseudocode program for this machine. After you have completed this, compare your pseudocode to the following:

```
Input state
Read colorCode
Case state
    1:
            Turn on machine
            If colorCode = 1 then
                Reject part
            End if
            IF colorCode = 2 then
                Accept part
            End if
            If ColorCode = 3 then
                Send part back for inspection
            End if
```

```
    2:
         Machine off
    3:
         Machine in Standby
End Case
```

Challenge

For this example, only the pseudocode will be provided. In this case, the pseudocode will be enough to implement a working program; however, a flowchart would really help a layperson, such as a customer, understand the process that we're attempting here. As such, as an exercise, try to render a flowchart that represents this mock program. As usual, there are no right or wrong answers, as long as your flow logically depicts the flow of the program. Once you do that, you can move on to the code implementation phase.

Code implementation

For this phase of the project, we need to take our design and turn it into working code. As we did before, try to implement the code on your own before looking at the working example.

Variables

The first bit of code we're going to implement is the variables. For this program to work, use the following code:

```
PROGRAM PLC_PRG
VAR
    input            : INT;
    machineState     : WSTRING;
    colorCode        : INT;
    machineAction    : WSTRING;

END_VAR
```

Once you have the variables in place, you can implement the core logic for the PLC program, which should look like this:

```
CASE input OF
    1:
         machineState := "Machine On";
         IF colorCode = 1 THEN
             machineAction := "Reject part";
         END_IF
         IF colorCode = 2 THEN
             machineAction := "Accept Part";
```

```
        END_IF
        IF colorCode = 3 THEN
            machineAction := "send back for inspection";
        END_IF
    2:
        machineState := "Machine off";
    3:
        machineState := "Machine Standby";
  END_CASE
```

The way this code works is simple; we have a simple state machine, and when the machine is on, it will read the `colorCode` variable. Depending on the color code, the machine will perform a certain action such as accepting, rejecting, or sending the part back for inspection.

For our first example, write 2 for `input` and 1 for `colorCode`. This should put the machine in an off state, and the color code should be ignored, like what can be seen in *Figure 11.11*:

Device.Application.PLC_PRG		
Expression	Type	Value
⬦ input	INT	2
⬦ machineState	WSTRING	"Machine off"
⬦ colorCode	INT	1
⬦ machineAction	WSTRING	""

Figure 11.11 – Machine off color code ignored

Next, put the machine in standby mode, and again the color code should be ignored.

Device.Application.PLC_PRG		
Expression	Type	Value
⬦ input	INT	3
⬦ machineState	WSTRING	"Machine Standby'
⬦ colorCode	INT	1
⬦ machineAction	WSTRING	""

Figure 11.12 – Machine standby mode and color code ignored

Finally, to see this machine in action, turn the machine on by writing 1 for `input`. When you do that, you should see the same output that is in *Figure 11.13*:

Device.Application.PLC_PRG		
Expression	Type	Value
input	INT	1
machineState	WSTRING	"Machine On"
colorCode	INT	1
machineAction	WSTRING	"Reject part"

Figure 11.13 – Machine rejects part

Now that the machine is in an on state, the `IF` statements in that case block are being read. Since we had 1 already written to the `colorCode` variable, the part was rejected. In other words, we simulated a color detection module reading a red sticker. Next, we need to verify that the machine can accept a part. So, what we're going to do is write the number 2 to the `machineState` variable. When you do that, you should be met with the output seen in *Figure 11.14*:

Device.Application.PLC_PRG		
Expression	Type	Value
input	INT	1
machineState	WSTRING	"Machine On"
colorCode	INT	2
machineAction	WSTRING	"Accept Part"

Figure 11.14 – Machine accepts part

Finally, we need to test the last condition and check to ensure that the machine can send a part back for inspection. To do this, write 3 to the `colorCode` variable. When you do, you should see the output that is in *Figure 11.15*:

Device.Application.PLC_PRG		
Expression	Type	Value
input	INT	1
machineState	WSTRING	"Machine On"
colorCode	INT	3
machineAction	WSTRING	"send back for inspection"

Figure 11.15 – Machine sends part back for inspection

Eureka! The machine performs as expected. Now, as stated before, this was using a simple color schema for simplicity. In real life, we would use something akin to an RGB indicator that sends back some type of string for the PLC to be read. This leads us to our final challenge!

Final challenge

Comparing a string of alphanumerical text is the same as comparing two numbers. For example, if we wanted to see if one WSTING variable is ABC123, we would use a block of code like this:

```
IF string_var = "ABC123" Then
    MachineAction := "some action";
END_IF
```

In this case, when string_var is ABC123, the machineAction variable will be set to "some action".

Typically, color modules may either return a numerical value or some type of code for a color, which may look like #FFF. For this challenge, modify the code so that when you enter #FFF for a WSTRING variable called hexColor, it will return white. Also, expand the program to take in some hex codes for red, blue, black, green, and yellow.

Summary

In this chapter, we explored IF and CASE statements, which are the backbone of flow control for any ST PLC programming environment. More importantly, we learned what flow control is and how to design programs that have many branches using a combination of pseudocode and flowcharts. In all, understanding how to control the flow of a program is one of the most basic and important skills a PLC programmer can have. As such, feel free to reread this chapter until you are comfortable with the material. Then, come back to it later when you're in the field working on a real-world project in ST.

Now, the IF statements in the CASE block are what are known as nested IF statements; essentially, that means that they are inside of another control structure. Also, it should be noted that the IF statements were structured in a less-than-ideal way. A more appropriate configuration would have been what's known as an IF-ELSE-IF configuration, which is a concept we haven't covered yet. This configuration, nested statements, and more are going to be explored in the next chapter!

Questions

1. What direction does a program always flow in?

2. When will an IF statement run?

3. Will an IF statement be executed if the condition evaluates to false?

4. What is a state machine?

5. What is a common way to implement state machines?

6. What is a CASE statement?

7. What is the CASE statement syntax?

8. Can you have an IF statement inside of a CASE statement?

9. What is the major difference between an IF and a CASE statement?

10. How can flowcharting help with flow control?

11. What is the difference between "less than or equal to" or just "less than" instructions?

12. What symbol is used to test for "not equals" in an IF statement?

13. What symbol is used to test for "greater than" in an IF statement?

14. What is the minimum amount of code needed for an IF statement?

15. What happens to the code in an IF statement when the expression evaluates to true?

12
Unlocking Advanced Control Statements

In the previous chapter, we took a deep dive into the world of flow control. If you found flow control confusing, you're in good company because many students and programmers inexperienced in ST are in the same boat and find the concept equally confusing. However, what we explored in the last chapter was just the tip of the iceberg when it comes to flow control. Flow control is much richer, and there are a few add-on commands that can be employed to really spruce up your PLC program.

Often, a block of code will need multiple conditions to be satisfied for the block of code that it contains to run. For example, if you think back to the beer example that we've explored multiple times throughout this book, not only does a person have to be at least 21 to buy a beer in the United States but they must also have enough money to cover it. This is an example of a situation that needs multiple conditions to be true for a person to buy beer. There is another caveat to this situation: what if the person is not in the United States? The legal drinking age varies wildly from country to country. As such, the past examples we've explored simply will not work. To create software that can accommodate situations like this, a more sophisticated variation of flow controls is needed.

Advanced flow control is used for so much more than just determining whether a person can buy beer. In this chapter, we're going to explore more advanced flow control and how to use it in the PLC world by exploring the following concepts:

- Nested control statements
- ELSE statements
- ELSIF statements
- Logical operators

Finally, to round out the chapter, we're going to modify the color sorter from the last chapter to account for shapes, too.

Technical requirements

This chapter requires a solid understanding of *Chapter 11*, Boolean logic and program design. If you are not comfortable with those concepts, you should review them before trying to tackle this chapter. In terms of technology, all that will be needed is a working copy of CODESYS. The code for the examples can be pulled down from the following URL:

`https://github.com/PacktPublishing/PLCs-for-Beginners`

As usual, you are encouraged to pull the code and explore it firsthand. It is also advisable to try to modify the code to gain a better understanding of the material that will be explored here.

Nested control statements

In the last chapter, we briefly looked at nested control statements. We had a CASE conditional with an IF statement inside it. When one control statement is nested inside another, they are typically referred to as either nested or embedded IF or CASE statements. A simple example of a nested IF statement is the beer and money analogy from before. Consider the following pseudocode:

```
Input age
Input money
If age >= 21 then
     If money >= 45 then
          Let them drink
```

For this example, two conditions must be true. First, the age of the person must be at least 21 years old. If so, they can move to the next conditional, which checks the amount of money they have. If they have at least $45, they can drink the beer. If any one of these conditionals is false, nothing will happen. In other words, they will not be able to drink! An easy way to think of these conditionals is as gatekeepers. To get to the glorious beer, a person would need to pass through the age gate and the money gate.

In terms of automation, nested control statements like these are very common. For example, suppose you're working on an industrial oven. For the oven to turn on, the door would have to be closed. Once the door is closed, the operator will be able to push a button to set the temperature of the oven. Consider the following pseudocode:

```
If doorState = "closed" then
     Turn off standby light
     If button1Pressed then
          Set temperature 100°F
     If button2Pressed then
          Set temperature 200°F
     If button3Pressed then
          Set temperature then 300°F
```

```
If doorState = "open" then
    Turn on standby light
    Set temperature to 0
```

In the case of this pseudocode, the door can be either in an open or closed state. If the door is closed, the temperature can be set to three different settings depending on the button that is pressed. If the door is in an open state, the temperature setting will be completely ignored and will have no effect.

To translate this into working code, we would need to implement the following variables:

```
PROGRAM PLC_PRG
VAR
    doorState : WSTRING;
    button1   : BOOL;
    button2   : BOOL;
    button3   : BOOL;
    temp      : INT;
    standby   : bool;
END_VAR
```

For this code, we have three BOOL-type variables that represent buttons on a control panel. Whichever button is set to TRUE will represent a button that is pressed. However, the heart and soul of this program is the doorState variable, which will determine whether the machine is in standby mode or not and whether the button will even be set. Finally, the temp variable will hold the oven's current set point temperature, and the standby variable will hold the state of the machine – that is, whether the machine is in standby mode or not.

Once you have those variables implemented, you can move on to implementing the logic for the program. This program's logic is long compared to previous examples but it is not complicated. The code is as follows:

```
IF doorState = "closed" THEN
    standby := FALSE;
    IF button1 = TRUE THEN
        temp := 100;
    END_IF
    IF button2 = TRUE THEN
        temp := 200;
    END_IF
    IF button3 = TRUE THEN
        temp := 300;
    END_IF
END_IF

IF doorState = "open" THEN
```

```
    standby := TRUE;
    temp := 0;
END_IF
```

Once you're done implementing the code, the first test you can do is to check the door's `open` functionality, which should set the temperature setpoint to 0 and put the machine in standby mode. To do this, set `doorState` to `open`. Once you do this, your output should match *Figure 12.1*:

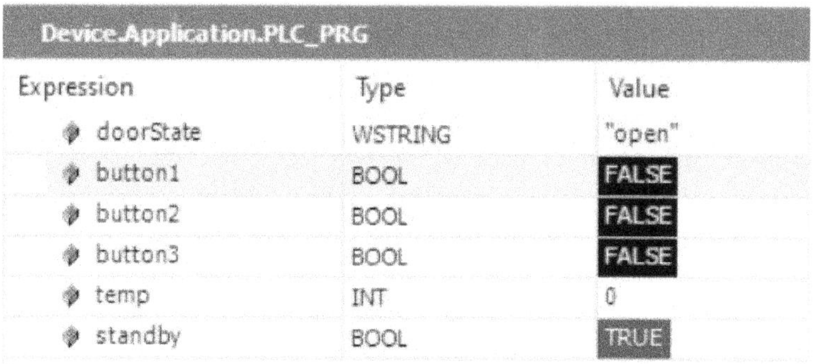

Device.Application.PLC_PRG		
Expression	Type	Value
doorState	WSTRING	"open"
button1	BOOL	FALSE
button2	BOOL	FALSE
button3	BOOL	FALSE
temp	INT	0
standby	BOOL	TRUE

Figure 12.1 – Machine standby mode

To explore how this code disables the button logic, try setting one of the `button` variables to `TRUE` and observe both the `temp` and `standby` variables' states.

Once you explore the `standby` functionality, you can move on to setting the temperature. For this functionality, set `doorState` to `closed` and set any of the `button` variables to `TRUE`. For this example, we are going to set the oven to its highest setting by setting `button3` to `TRUE`. Once you do that, you should be met with *Figure 12.2*:

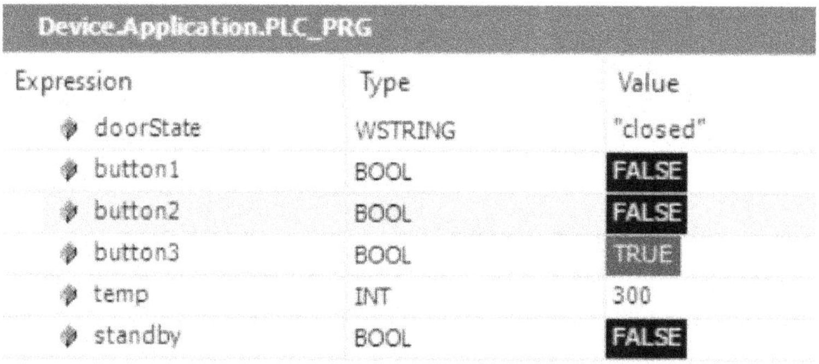

Device.Application.PLC_PRG		
Expression	Type	Value
doorState	WSTRING	"closed"
button1	BOOL	FALSE
button2	BOOL	FALSE
button3	BOOL	TRUE
temp	INT	300
standby	BOOL	FALSE

Figure 12.2 – Door closed and high temp mode

For an experiment, test the other settings by turning the current temperature button to `FALSE` and any of the other buttons to `TRUE`. Observe the temperature output. Also, set the door state to `open` to observe the temperature and machine state.

Challenge

Is there a bug in this code? Learning how to verify embedded control statements is a pivotal yet hard-to-acquire skill. Before you move on to exploring the next section, take a moment to explore what happens when you have multiple buttons pressed at the same time. Can you spot anything out of the ordinary? Once you feel satisfied with your answer, you can move on to the almighty ELSE statement!

ELSE statements

Oftentimes, we'll need a default block of code to run if the IF statement does not evaluate to TRUE. To accomplish this, the ELSE statement is employed. The ELSE statement is like a catcher in a baseball game. If the batter misses the ball, it's the catcher's job to catch the ball and throw it back to the pitcher. Consider the following example:

```
If motor1 = selected then
     turn on motor1
     turn off motor2
Else
     turn on motor2
     turn off motor1
```

For this example, if motor1 is selected, that motor will turn on. If motor1 is not selected, then motor2 will turn on. Again, the ELSE block is like the catcher; if motor1 is not selected, it will catch that and turn on motor2. To see this in action, implement the following variables:

```
PROGRAM PLC_PRG
VAR
     selectMotor1            : BOOL;
     isMotor1Selected        : BOOL;
     isMotor2Selected        : BOOL;
END_VAR
```

This example uses selectMotor1 as the main input that will determine the state of isMotor1Selected or isMotor2Selected. The logic for the program will look like the following:

```
IF selectMotor1 = TRUE THEN
     isMotor1Selected := TRUE;
     ismotor2Selected := FALSE;
ELSE
     isMotor1Selected := FALSE;
     ismotor2Selected := TRUE;
END_IF
```

When the code is executed, `selectMotor1` will default to `FALSE` and will render *Figure 12.3*:

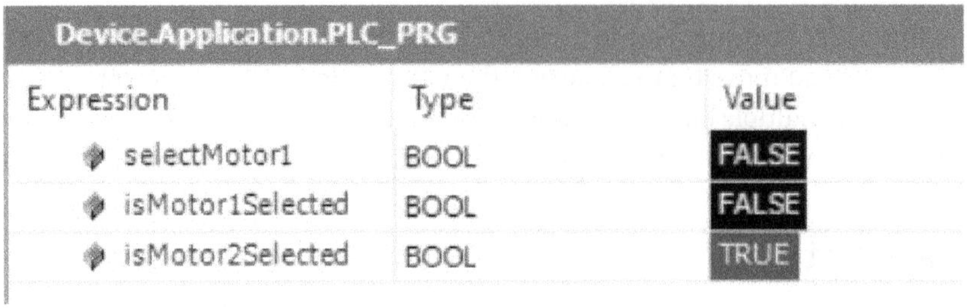

Device.Application.PLC_PRG		
Expression	Type	Value
⬧ selectMotor1	BOOL	FALSE
⬧ isMotor1Selected	BOOL	FALSE
⬧ isMotor2Selected	BOOL	TRUE

Figure 12.3 – The selectMotor1 variable default of FALSE

When the `selectMotor1` variable is toggled to `TRUE`, it will produce *Figure 12.4*:

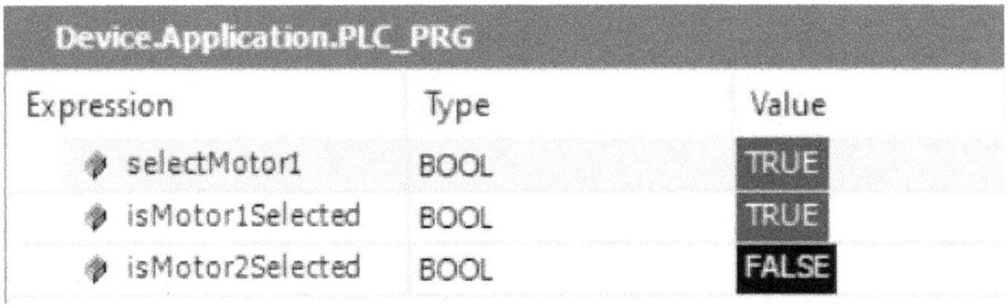

Device.Application.PLC_PRG		
Expression	Type	Value
⬧ selectMotor1	BOOL	TRUE
⬧ isMotor1Selected	BOOL	TRUE
⬧ isMotor2Selected	BOOL	FALSE

Figure 12.4 – The selectMotor1 variable set to TRUE

When `selectMotor1` is toggled to `TRUE`, that `IF` statement will run that code block. This will result in `motor1` turning on and `motor2` turning off.

There is a caveat to the `ELSE` statement. Since the `ELSE` statement is like a default statement, you can only have one per `IF` statement. As such, it is important to always remember the saying "One `IF` statement, one `ELSE` statement!"

> **Note**
>
> The `ELSE` statement can also be used in the same way as a `CASE` statement. All of the rules that apply to the `IF` statement also apply to the `CASE` statement.

`ELSE` is a very important addition to augmenting `IF` statements; however, it's not the only addition that can be used. In the next section, we're going to combine multiple `IF` statements together to form one large cohesive unit!

ELSIF statements

Typically, in automation, a program will usually have multiple code blocks that will run under different situations. An example of this would be the state machines we explored with CASE in the last chapter. Now, CASE statements are excellent ways to build state machines and orchestrate multiple code blocks; however, they are not the only way. Another very common way of constructing something akin to a state machine is with the ELSIF command.

When coupled with an IF statement, ELSIF will create a cohesive conditional block that will accommodate multiple conditions. For example, consider our beer discussion. If a person is in the United States, they must be at least 21 years old to purchase beer. On the other hand, if a person is in Mexico, they must only be 18 years old. If we needed a program that could handle this, we could use something like the following pseudocode:

```
Input age
If country = "usa" then
    If age >= 21 then
        Let them drink!
Else_If country = "mexico" then
    If age >= 18 then
        Let them drink!
```

To see how this code would work, let's explore a working example! The first thing we need to do is declare the necessary variables as usual:

```
PROGRAM PLC_PRG
VAR
    country : WSTRING;
    buyBeer : BOOL;
    age     : INT;
END_VAR
```

For this code, we have a country variable that will determine the drinking age. The next variable is a Boolean that will determine whether we can buy beer or not. It will be set to TRUE if we can buy beer; otherwise, it will be set to FALSE. Finally, we have an age variable that can be used to take in a user's age.

The logic for this program will look like the following:

```
IF country = "usa" THEN
    IF age >= 21 THEN
        buyBeer := TRUE;
    ELSE
        buyBeer := FALSE;
    END_IF
```

```
ELSIF country = "mexico" THEN
    IF age >= 18 THEN
        buyBeer := TRUE;
    ELSE
        buyBeer := FALSE;
    END_IF
END_IF
```

When you run the program, enter usa for country and 28 for age. Once done, you should see the same output that's in *Figure 12.5*:

Device.Application.PLC_PRG		
Expression	Type	Value
⬧ country	WSTRING	"usa"
⬧ buyBeer	BOOL	TRUE
⬧ age	INT	28

Figure 12.5 – Over 21 in USA output

Now, if we were to input 18 for age and keep the country variable set to usa, we would get the following:

Device.Application.PLC_PRG		
Expression	Type	Value
⬧ country	WSTRING	"usa"
⬧ buyBeer	BOOL	FALSE
⬧ age	INT	18

Figure 12.6 – 18 in the USA output

Notice that since we are in the USA, we can't buy beer. Now, if we change country to mexico, we'll get *Figure 12.7*:

Device.Application.PLC_PRG		
Expression	Type	Value
⬧ country	WSTRING	"mexico"
⬧ buyBeer	BOOL	TRUE
⬧ age	INT	18

Figure 12.7 – 18 in Mexico output

In this case, the buyBeer variable toggles to TRUE! This is because changing which of the IF statements we are in will affect the age check. In other words, the country variable is the main gatekeeper to the drinking age. This is a very important concept to remember, especially when troubleshooting. Many times, it's easy to forget which block is running. As a result, you can be troubleshooting a block that is working as it should be!

Our beer-buying program is a great example of how an IF-ELSIF statement works. However, it's not exactly how one would use it in real life. A common IF-ELSIF is something akin to a temperature monitoring application. Typically, for an application like this, a programmer will usually program a series of IF-ELSIF statements with the intention of performing certain functions, such as turning on warning lights or locking/unlocking doors when a certain temperature is detected. Consider the following pseudocode design:

```
Read temp
IF temp >= 300 then
     Turn on red light
     Turn off yellow light
     Turn off green light
ELSIF temp >= 200 then
     Turn on yellow light
     Turn off green light
     Turn off red light
ELSIF temp >= 100 then
     Turn on green light
     Turn off yellow light
     Turn off red light
```

This design will essentially read the temperature and turn off the unnecessary lights while turning on the correct light. To see this program implemented, set up the following variables:

```
PROGRAM PLC_PRG
VAR
     temp            : INT;
     redLight        : BOOL;
     yellowLight     : BOOL;
     greenLight      : BOOL;
END_VAR
```

In this case, we have three light variables that are Boolean data types. To represent an on state for the light, the variable will be set to TRUE. Whether a light is on will depend on the temp variable, which will hold a simulated temperature reading. As for the logic, we can use the following:

```
IF temp >= 300 THEN
     redLight     := TRUE;
     greenLight   := FALSE;
```

```
     yellowLight := FALSE;
ELSIF temp >= 200 THEN
     redLight    := FALSE;
     greenLight  := FALSE;
     yellowLight := TRUE;
ELSIF temp >= 100 THEN
     redLight    := FALSE;
     greenLight  := True;
     yellowLight := FALSE;
END_IF
```

The temperature will essentially dictate which code block to run and, by extension, which light will be turned on. When the program is run and the `temp` variable is set to `100`, we should get the output that can be seen in *Figure 12.8*. As can be seen in *Figure 12.8*, when the temperature is greater than `100` but still less than `200`, the green light will turn on while all the other lights will be put into an off state:

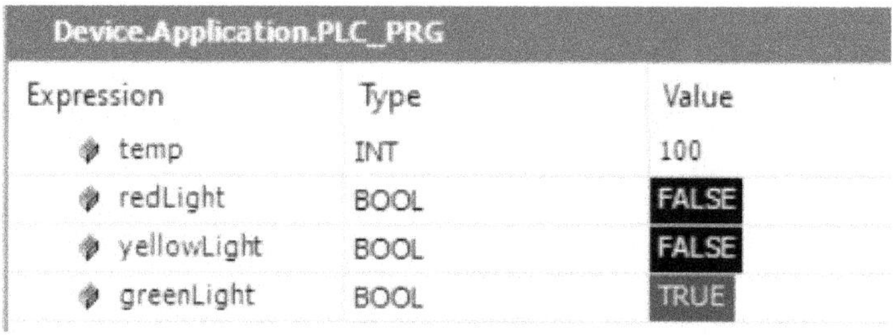

Device.Application.PLC_PRG		
Expression	Type	Value
temp	INT	100
redLight	BOOL	FALSE
yellowLight	BOOL	FALSE
greenLight	BOOL	TRUE

Figure 12.8 – Green light on

Now, the next thing to test is the functionality of the yellow light. To do this, set the temperature to a value between `200` and `300`. In the case of *Figure 12.9*, `250` will be used:

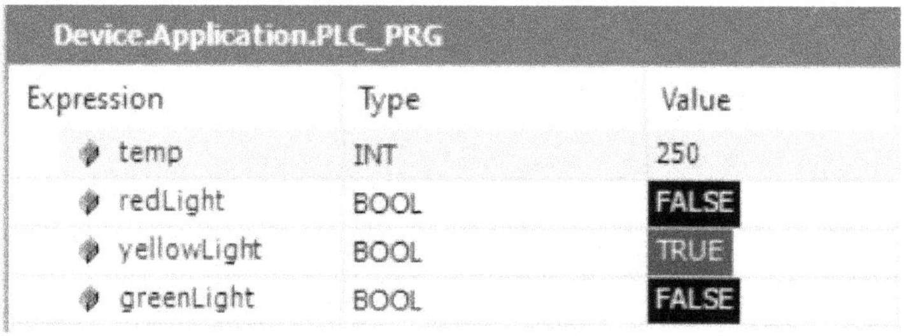

Device.Application.PLC_PRG		
Expression	Type	Value
temp	INT	250
redLight	BOOL	FALSE
yellowLight	BOOL	TRUE
greenLight	BOOL	FALSE

Figure 12.9 – Yellow light on

As can be seen, the yellow light is switched on and the green light that was on before was turned off. The last thing to test is the red light. For this, we will set the `temp` variable to `311`:

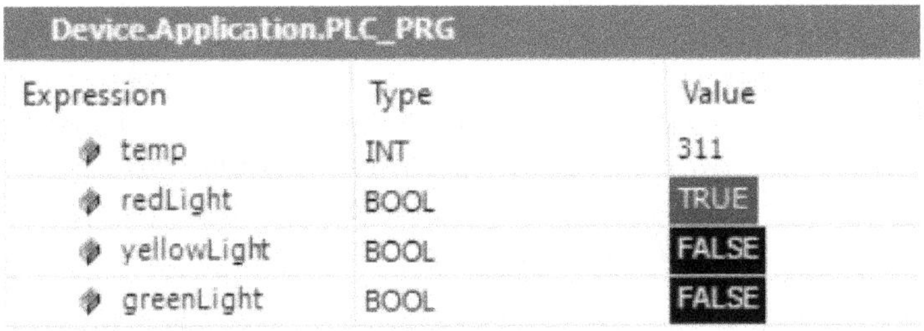

Figure 12.10 – Red light on

As can be seen, the red light was switched on while all the others were switched off.

An IF-ELSIF block is like a race condition. The first IF or ELSIF statement to evaluate to TRUE will "win" and that code will execute. As such, order matters with these blocks, especially when one is working with number comparisons, kind of like the example we just explored.

Consider the following example:

```
Read temp
IT temp >= 100 then
     Turn on green light
ELSEIF temp >= 200 then
     Turn on yellow light
ELSEIF temp >= 300 then
     Turn on red light
```

Working off the assumption that the first block to evaluate to TRUE will "win" or execute, what do you think will happen when the temperature reaches 201°F? If you answered that the green light will turn on, you would be right. Again, this is because 201 is greater than 100. Since the first IF block will evaluate to TRUE when it reads a temperature greater than 100, such as 201, it will turn on the green light, and then the IF-ELSIF block will terminate.

Challenge

Here is a challenge to see how the race condition works. Take the working code for the original temperature program and modify it so that the program checks for temperature in ascending order. In other words, convert the pseudocode from the last explanation into working code. Once you implement that code, play with your input values and see how the program behaves. Chances are, you will see the bug.

Now that we have explored the IF, ELSE, and ELSIF statements, we need to switch our attention to creating more complex logical statements. That is, we need to explore what happens when we start mixing multiple TRUE or FALSE conditions into the same control statement!

Logical operators

In automation, it is very common for multiple conditions to be met before a system can perform a task. For example, in some cases, we may need to ensure that all the safety switches are open and the start button is engaged before the machine can start up. In other cases, we may need two motor drives to be on before the machine can try to move. To accomplish this, we need to look at more complex logical expressions.

So far, we have seen some code (nested `IF` statements) that will behave in such a way that multiple conditions need to be met before a certain task can be carried out. Nested states are suitable when some code needs to be executed based on a single condition while other code requires an extra condition to be satisfied. However, for many cases where all the code needs to be executed when multiple conditions are met, the nested statements can be cumbersome and bloat the code. This is mainly because all of our `IF` or `ELSIF` statements have only been used as single logical conditions to determine whether the block of code should run or not. As stated before, this is not optimal. As such, enter the world of complex logic statements.

What we're going to call a complex logical statement is a control statement that will use more than one condition to evaluate to `TRUE`. In essence, these conditionals will be applied to Boolean equations. If you haven't read *Chapter 6*, you may want to explore that chapter now.

The main workhorse that we're going to use for this section is truth tables. For control statements like the ones we're about to explore, a truth table can often be a handy troubleshooting tool and really help speed up the development process. As such, for our first example, let's revisit our beer and money program.

If you remember, in the original beer and money program, we used a nested `IF` statement to determine whether the person was old enough to buy beer and had the necessary money to buy it. That example worked fine; however, it wasn't the most optimal solution. In the IEC 61131-3 standard, complex logical statements are produced using the following keywords to determine their corresponding logical operations:

- `AND`
- `OR`
- `XOR`
- `NOT`

The simplest way to demonstrate these operations is to see one in action. Let's modify our beer and money program to use the `AND` statement. To do this, let's start off by looking at some pseudocode to give us a quality roadmap to the real program:

```
Input age
Input money
If money >= 45 AND age >=21 then
```

```
        Let them drink heavily
   Else
        Throw a shoe at them
```

The magic in this program resides in the AND keyword. Essentially, we can use the following truth table to determine whether we can buy beer or not:

Money >= 45	Age >= 21	Can they drink heavily?
Yes	Yes	Yes
Yes	No	No
No	Yes	No
No	No	No

Table 12.1 – Can they drink?

In this case, the only way they can drink is if they have more than $45 and they are at least 21. To see this behavior, let's implement the program.

For this program, we're going to use the following variables:

```
PROGRAM PLC_PRG
VAR
    money : INT;
    age   : INT;
    msg   : WSTRING;
END_VAR
```

These variables are self-explanatory, so once you have those implemented, use the following code for the program's main logic:

```
IF money >= 45 AND age >= 21 THEN
    msg := "Let them drink heavily";
ELSE
    msg := "Throw a shoe at them";
END_IF
```

Once you have all that implemented, run the program and input 50 for money and 22 for age. When you write those variables, you should be met with *Figure 12.11*:

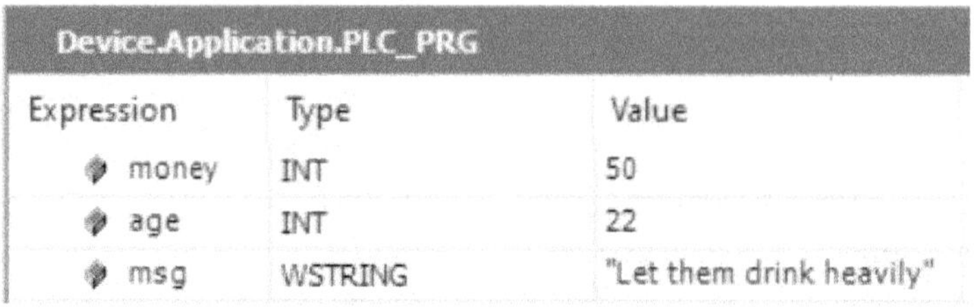

Device.Application.PLC_PRG		
Expression	Type	Value
⬧ money	INT	50
⬧ age	INT	22
⬧ msg	WSTRING	"Let them drink heavily"

Figure 12.11 – The AND statement for the beer program, age 22

As can be seen, since we have more than $45 and we are older than 21, the program says we can drink heavily. Now, let's change our age to 19. When you do that, you should be met with *Figure 12.12*:

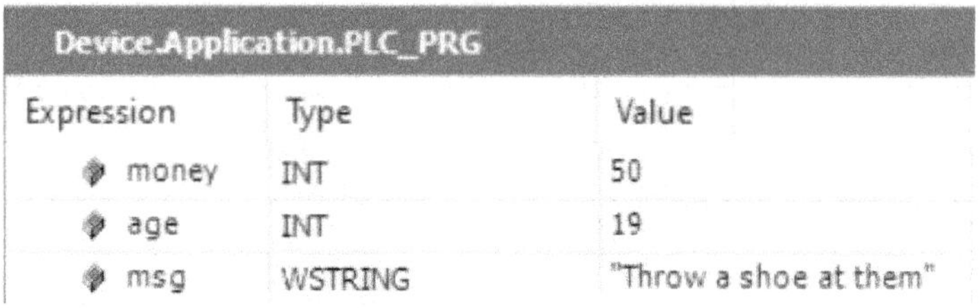

Device.Application.PLC_PRG		
Expression	Type	Value
⬧ money	INT	50
⬧ age	INT	19
⬧ msg	WSTRING	"Throw a shoe at them"

Figure 12.12 – The AND statement for the beer program, age 19

As can be seen, when we enter 19 for age, our program will throw a shoe at the patron. This is because if we consider the truth table, for the patron to buy beer, they must be at least 21 and have at least $45. In the case of *Figure 12.12*, one of those conditions is FALSE, and as such, the whole statement will evaluate to FALSE.

The AND keyword is commonly used; however, it's not the only operator that can be used. In the next example, we're going to explore the OR operator.

Exploring the OR operator

As we explored in *Chapter 6*, for the OR operator to produce a TRUE or on state, at least one of the expressions needs to evaluate to TRUE. An example application for this would be a machine that has redundant power sources and only needs one to turn on. A pseudocode design for a program like this would be as follows:

```
Read powersource1
Read powersource2
If powersource1 == on or powersource2 == on then
     Turn on machine
Else
     Turn off machine
End if
```

If we were to turn this into working code, it would look like the following:

```
PROGRAM PLC_PRG
VAR
     powersource1 : BOOL;
     powersource2 : BOOL;
     machineState : BOOL := FALSE;
END_VAR
```

In this example, we have three Boolean variables. The two `powersource` variables will serve as input variables. If either of these variables is TRUE, then the machine state will toggle to TRUE. The logic for this program is as follows:

```
IF powersource1 = TRUE OR powersource2 = TRUE THEN
     machineState := TRUE;
ELSE
     machineState := FALSE;
END_IF
```

The logic for this program is quite simple. The OR statement will trigger the `machineState` variable to toggle to TRUE when either of the `powersource` variables is on. As such, once you run the program and set either or both of the `powersource` variables to TRUE, the `machineState` variable will change. Consider *Figure 12.13*:

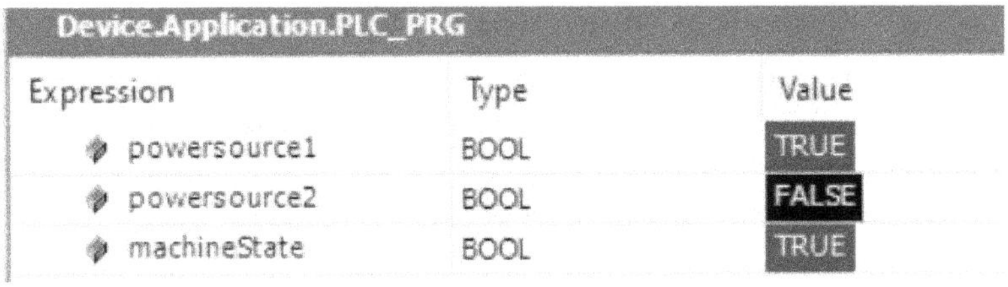

Device.Application.PLC_PRG		
Expression	Type	Value
powersource1	BOOL	TRUE
powersource2	BOOL	FALSE
machineState	BOOL	TRUE

Figure 12.13 – Powersource1 on

As can be seen, the `machineState` variable is TRUE. Next, set `powersource1` to FALSE and `powersource2` to TRUE. The output should match *Figure 12.14*:

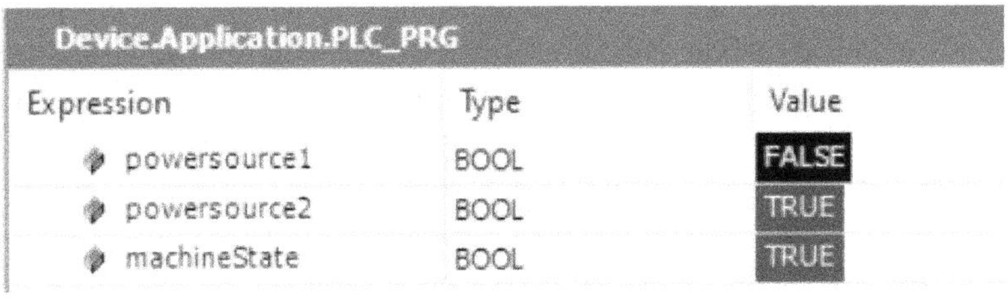

Device.Application.PLC_PRG		
Expression	Type	Value
powersource1	BOOL	FALSE
powersource2	BOOL	TRUE
machineState	BOOL	TRUE

Figure 12.14 – Powersource2 on

Again, with only one `powersource` variable set to on, `machineState` is still TRUE. Next, we're going to turn on both power sources:

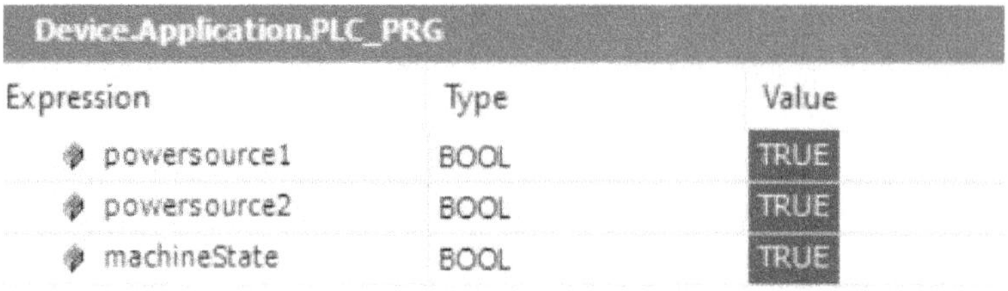

Device.Application.PLC_PRG		
Expression	Type	Value
✦ powersource1	BOOL	TRUE
✦ powersource2	BOOL	TRUE
✦ machineState	BOOL	TRUE

Figure 12.15 – Both power sources on

Again, *Figure 12.15* shows that with both power sources on, the machine state does not change to FALSE. To see a change in the machine state, you will need to turn both `powersource` variables off:

Device.Application.PLC_PRG		
Expression	Type	Value
✦ powersource1	BOOL	FALSE
✦ powersource2	BOOL	FALSE
✦ machineState	BOOL	FALSE

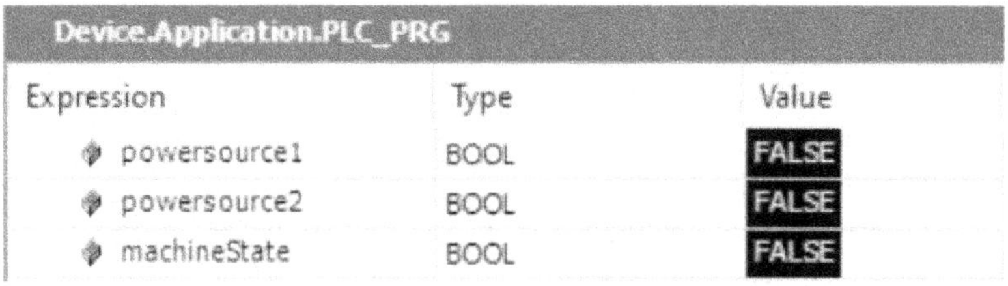

Figure 12.16 – Both powersource variables off

To set `machineState` to FALSE, both `powersource` variables must be set to FALSE. This is the nature of the OR operator. As long as at least one of the conditions evaluates to TRUE, the code block will run. The next operator that we need to explore is the XOR operator.

Exploring the XOR operator

Of all the logical operators, XOR is arguably used the least. Since this operator will only result in a TRUE condition when one and only one of the inputs is TRUE, there is often little use for the operator. However, as with anything else, there is a time and a place for it. Consider the power source example we explored with the OR statement. Sometimes in automation, you will only want one power source to power the machine at a time. Depending on how the machine is designed, two power supplies being on at the same time can result in damage. This is where the XOR statement comes into play. For this example, we're going to rework the OR operator example and create a program that will only power on when one power supply is on. As such, we're going to keep the same variables we used in the last example and modify the logic to match the following:

```
IF powersource1 = TRUE XOR powersource2 = TRUE THEN
    machineState := TRUE;
ELSE
    machineState := FALSE;
END_IF
```

When you run the code, set one of the `powersource` variables to TRUE and you should be met with *Figure 12.17*:

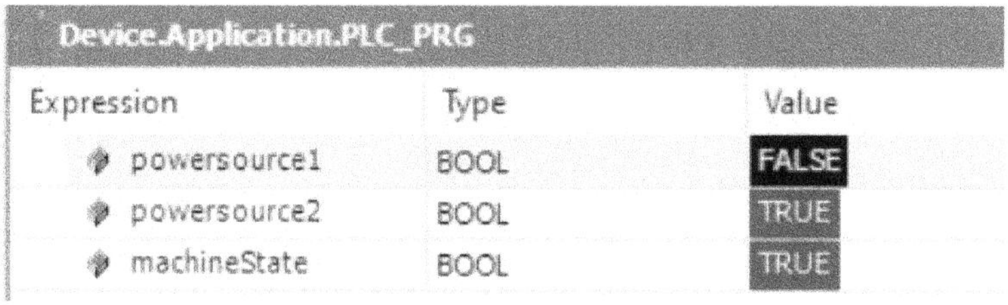

Figure 12.17 – One powersource variable on

As can be seen, the machine state goes to TRUE or on. Now, turn both `powersource` variables on. When you do, you should be met with *Figure 12.18*:

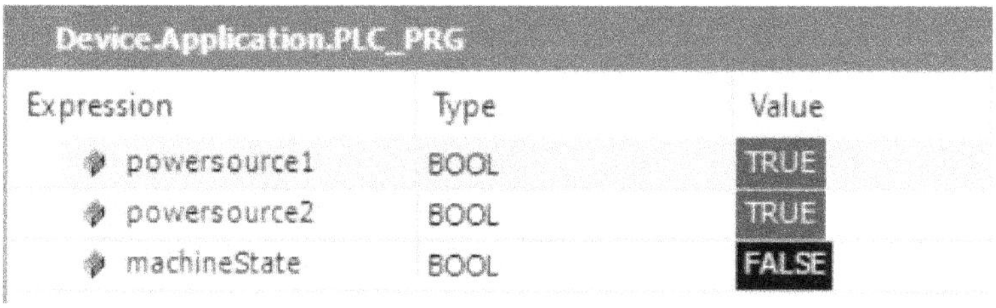

Figure 12.18 – Both powersource variables on

As is depicted in the figure, the machine state is FALSE or off.

Now that we've explored the operators that take two inputs, we need to switch our attention to the NOT operator.

Exploring the NOT operator

The last logic operator that we need to explore is the NOT operator. The NOT operator is a very simple operator that merely inverts the logical expression. In other words, it will turn TRUE to FALSE and FALSE to TRUE. This operator is handy when you need to perform a check on the same variable and require one condition to be TRUE and one to be FALSE. Consider the following pseudocode:

```
Var = False
If NOT var == True
      Var is false but will read as true
Else if NOT var == False
      Var is true but will read as false
```

Translating this into real code will yield the following result.

In terms of variables, the code should look like the following:

```
PROGRAM PLC_PRG
VAR
      variable : BOOL;
      msg      : WSTRING;
END_VAR
```

In terms of the code, the logic should be the following:

```
IF (NOT variable) = FALSE THEN
      msg := "variable is set to True";
ELSE
      msg := "variable is set to False";
END_IF
```

When you run the code and set variable to TRUE, you should get what's in *Figure 12.19*:

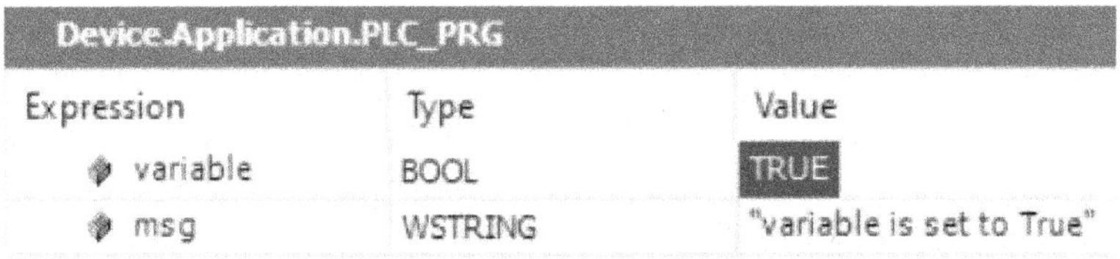

Device.Application.PLC_PRG		
Expression	Type	Value
◈ variable	BOOL	TRUE
◈ msg	WSTRING	"variable is set to True"

Figure 12.19 – A NOT example (True input)

If you look at the IF statement, it will only run when the expression equals FALSE; however, our input variable is set to TRUE. If you notice, the output the block ran. This is because the NOT statement inverted the expression and set the overall conditional to TRUE, which allowed the block to run.

Now that you've finished this section, you should have a solid understanding of complex logical expressions. The expressions we've explored were essentially one or two inputs. However, you can string as many operators as you need together to simulate more inputs. In all, you should now have enough understanding to finish the chapter with our final project.

Final project

For our final project, we're going to make a derivative of the color sorter we made in the last project; however, this time, there will be a catch. Our program is going to sort not just colors but shapes as well. Therefore, the first order of business we're going to do is lay out a list of requirements.

Requirements

For this project, we're going to need to meet the following requirements:

1. Red shapes will either be squares or rectangles:

    If a shape is detected as a square or rectangle, it will go down the square line, else it will be rejected

2. Green shapes will always be stars:

    If a shape is a star, it will go down the star line, else it will go down the reject line

3. Blue shapes will always be triangles:

    If a shape is a triangle, it will go down the triangle line, else it will go down the reject line

4. Any other color will be rejected.

With these requirements, we should have enough information to build our program. As with all of our other examples, try to take a moment and implement the program based on the requirements before proceeding.

Variables

The variables for this program will be as follows:

```
PROGRAM PLC_PRG
VAR
     color  : WSTRING;
     shape  : WSTRING;
      line  : WSTRING;
END_VAR
```

In short, the `line` variable will hold the line the part is going down, `shape` will hold the `shape` type, and finally, `color` will hold the color of the part that the machine detects. Once these variables are implemented, we can write the logic.

Color and shape sorter logic

The core logic for the program will be as follows:

```
IF color = "red" THEN
     IF shape = "square" OR shape = "rectangle" THEN
          line := "square line";
     ELSE
          line := "reject";
     END_IF
ELSIF color = "green" THEN
     IF shape = "star" THEN
          line := "star line";
     ELSE
          line := "reject";
     END_IF
ELSIF color = "blue" THEN
     IF shape = "triangle" THEN
          line := "triangle line";
     ELSE
          line := "reject";
     END_IF
ELSE
     line := "rejected";
END_IF
```

As can be seen, this program is a series of nested `IF-ELSIF` statements. The first check will determine what the color is. The reasoning behind this is that the requirements group the shapes by color. In other words, the color kind of acts as a key to the shape category. As such, we have three checks for red, green, and blue. If a shape is none of those colors, it gets sorted to the reject line.

Nested inside the color checks is the shape check. This logic will ensure that the shape is the correct color. For example, we don't want a green square going into production. Therefore, if one of the colors we're checking for is detected, a secondary check will be conducted to determine the shape of the part. If the part is not recognized for that color, it will also be rejected (the `ELSE` statement logic).

Pay attention to the nested logic in the red block. Notice that there is an `OR` statement there. This is because both squares and rectangles can be red. Therefore, since both shapes can be red, we need to check to see whether the shape is one or the other. This is an important detail in the requirements, and this is a requirement that can trip up a lot of inexperienced programmers. Since they see the word "and" in the requirements, they will try to use the `AND` operator in the code.

In all, once the code has been implemented, test it out!

Testing conditions

The first condition we're going to test is for red colors. Enter square for shape and red for color:

Device.Application.PLC_PRG		
Expression	Type	Value
color	WSTRING	"red"
shape	WSTRING	"square"
line	WSTRING	"square line"

Figure 12.20 – Red square check

As can be seen, the part went down the correct line. Now that we know the system can detect a red square, let's make sure it can detect a red rectangle:

Device.Application.PLC_PRG		
Expression	Type	Value
color	WSTRING	"red"
shape	WSTRING	"rectangle"
line	WSTRING	"square line"

Figure 12.21 – Red rectangle check

Figure 12.21 shows that the part is still going down the correct line. We can now move on to testing for a blue rectangle. To do this, simply change the color from red to blue:

Device.Application.PLC_PRG		
Expression	Type	Value
color	WSTRING	"blue"
shape	WSTRING	"rectangle"
line	WSTRING	"reject"

Figure 12.22 – Blue rectangle check

Again, the program is working, as this part was rejected. With the squares out of the way, let's check the green logic functionality. For this test, set shape to triangle and color to green:

Device.Application.PLC_PRG		
Expression	Type	Value
◈ color	WSTRING	"green"
◈ shape	WSTRING	"triangle"
◈ line	WSTRING	"reject"

Figure 12.23 – Green triangle check

As expected, the part was rejected because triangles are always blue, and only stars can be green. So, to confirm the program will accept a green star, let's change the shape variable to star:

Device.Application.PLC_PRG		
Expression	Type	Value
◈ color	WSTRING	"green"
◈ shape	WSTRING	"star"
◈ line	WSTRING	"star line"

Figure 12.24 – Green star check

Figure 12.24 shows that the green block is working as expected, as it rejected a green triangle but put a green star down the correct line. We can now test out the functionality for the blue logic block. So, set color to blue and shape to hexagon:

Device.Application.PLC_PRG		
Expression	Type	Value
◈ color	WSTRING	"blue"
◈ shape	WSTRING	"hexagon"
◈ line	WSTRING	"reject"

Figure 12.25 – Blue hexagon check

This test was a little bit of a curve ball because hexagon is not mentioned anywhere in the program. This means that since we detected a blue hexagon, the logic in the blue IF block still rejected the part because only triangles can be blue. In other words, the program is performing as expected. To put the final touches on the blue logic block, set shape to triangle:

Device.Application.PLC_PRG		
Expression	Type	Value
⚙ color	WSTRING	"blue"
⚙ shape	WSTRING	"triangle"
⚙ line	WSTRING	"triangle line"

Figure 12.26 – Blue triangle check

Figure 12.26 shows that the part went down the expected line. Now, there is still one last condition to check. To ensure our program is working as expected, we need to ensure that any part, regardless of shape, is rejected if the color is not red, green, or blue. The ELSE statement at the end of the program should handle this condition. Simply change the color variable to any other color, such as black:

Device.Application.PLC_PRG		
Expression	Type	Value
⚙ color	WSTRING	"black"
⚙ shape	WSTRING	"triangle"
⚙ line	WSTRING	"rejected"

Figure 12.27 – Black color check

As can be seen in *Figure 12.27*, though the part is still a triangle, it was automatically rejected because it was black and not red, green, or blue. Overall, we can mark this program as a major success because it works!

Summary

This chapter has been a fast-paced introduction to complex flow control statements. The major takeaways from this chapter should be how to use nested control statements, complex logical expressions, and ELSIF/ELSE statements. As we progress through the book, we will use these principles more. If you do not yet understand these concepts, please read the chapter again to ensure you understand the material. With that, we're going to move on to our next chapter and learn about implementing loops!

Further reading

Fernhill SCADA – IF statements: https://www.fernhillsoftware.com/help/iec-61131/structured-text/st-if.html#:~:text=The%20IEC%2061131%2D3%20ST,ELSEIF%20condition%20THEN%20Statement%3B%20

Questions

1. What is the difference between AND and OR operators?

2. What is the NOT operator used for?

3. What is a nested IF statement?

4. What' the difference between ELSE and ELSIF statements?

5. Can a CASE statement use an ELSE statement?

6. Can an ELSE statement accept a logical expression?

7. What is the truth table for an XOR statement?

Implementing Tight Loops

Almost every program, regardless of whether it is a PLC program or a normal application, will loop in some fashion. If a PLC program did not loop, it would be relatively worthless because unless the operators had near-perfect timing for starting an operation, the program would almost immediately stop. Even if the operator did manage to kickstart a run, as soon as the program reached its final command, it would stop. Without some type of loop, a PLC program would be relatively useless.

PLCs will often implement a loop that will iterate over a program and prevent the PLC program from needing to be restarted. However, programmers will often need loops to be a bit more specific. More specifically, a programmer will often need to create a loop to loop over a smaller portion of code in the PLC program until a condition is met.

In programming, there are many different types of loops. Each type of loop will have its own applications and optimization for certain tasks. Understanding which type of loop to implement is paramount to writing quality PLC code. Using the wrong type of loop can result in program bugs, a bloated, inefficient code base, or in a worst-case scenario, create situations that can put the machine or operators in danger. It goes without saying that knowing how to design quality loops is a must for any PLC programmer.

Using quality loops is vital to the successful implementation of a PLC program. This chapter is going to explore the types of loops that are governed by the IEC 61131-3 standard and how to properly use them. To do this, we're going to explore the following concepts:

- Types of loops
- Loops with a flowchart and pseudocode
- FOR loops
- WHILE loops
- REPEAT loops
- Nested loops

To round out this chapter, we're going to create a factory assembly line with three states. One line state will be used to build a certain number of parts, another line state will build at least one part, and finally, the third line state will be used to build parts only under certain conditions.

Technical requirements

The code for this chapter can be found at the following link: `https://github.com/PacktPublishing/PLCs-for-Beginners`

This chapter will also require flowchart designs, so you can use `draw.io` to draw the flowcharts at the following URL, or use any other drawing method you like: `https://app.diagrams.net/`

Exploring the different types of loops

The IEC 61131-3 standard governs three different types of loops. The three types of governed loops are as follows:

- Counter loops
- Precheck loops
- Postcheck loops

If you've programmed in a language such as C++, Java, or C# before, you're probably already familiar with these types of loops. In practice, the loops will behave the same as in ST. Regardless, we're going to explore what they are and how they work.

Counter loop

A counter loop is straightforward. A counter loop will loop over a block of code a certain number of times. This means that a counter loop will start with a specific number, increment, and then terminate when a numerical condition is met. For example, a counter loop may use a counter variable that starts at 1 and will loop for as long as the counter variable is less than, say, 20.

Most programming languages use what's called a `FOR` loop to implement a counter loop, and ST is no different. Of all the loops you're going to use throughout your career, you'll use the `FOR` loop the most. Applications for the `FOR` loop can be as follows:

- Looping through arrays
- Running a process to create a certain number of parts
- Sending pulses to a motor

There is a near-infinite number of applications for the FOR loop. However, these are just a few commonly used applications for a PLC programmer.

Let's now explore another commonly used but not as popular loop that's called a precheck loop.

Precheck loops

Though not as common as a counter loop, precheck loops are probably the second most common type of loop to implement. A precheck loop will check some type of logical expression, similar to an IF statement. As long as that condition is TRUE, the loop will loop over a block of code.

When the program first detects the loop, it will evaluate the condition. This means that you are not guaranteed that the code inside the loop will run. This is very important and can confuse many inexperienced programmers. It is a common pitfall for many new programmers to think they are guaranteed at least one iteration with this loop when, in reality, they are not.

Much like the counter loop, most programming languages usually use the WHILE keyword to signal the precheck loop. If you have programmed in a traditional programming language, you'll notice the ST syntax is similar to that of a language such as C++ or Java with extra keywords.

With that, we can move on to post check loops.

Post check loops

Post check loops are not available in every programming language. For example, the programming language Python does not support this type of loop; however, many other popular languages do. This type of loop is similar to the precheck loop with one caveat. A post check loop will evaluate a logical condition, much like a precheck loop; however, the difference is that the post check loop will evaluate the condition at the end of the loop. This means that you are guaranteed at least one iteration of the code inside the loop.

In terms of applications, the post check loop can be leveraged to great success in PLC programming. This loop is handy for applications such as automated saws, welding machines, or drills. For example, suppose you're working on an automated welding machine. A machine may create a quality weld on the first run, or the machine may have to move over the part multiple times to ensure a quality weld. For applications such as these, the post check loop is an ideal loop. In this case, a post check loop will guarantee that you will get at least one welding pass over a part, but if the part has to be reworked, it can iterate again and instruct the machine to do so.

Most languages call a post check loop a DO or DO UNTIL loop. However, unlike the other types of loops we've explored, ST uses a different set of keywords for the loop. In ST, a post check loop is called a REPEAT loop. Remembering this piece of information is vital to a PLC programmer, as it is not that uncommon for a developer who is experienced with a traditional language to try and use a similar syntax.

> **Note**
> It is important to remember that the DO keyword does exist in ST. However, the DO keyword is used in conjunction with the WHILE and FOR loops and does not trigger a loop on its own.

Infinite loops

Normally, we want a loop to eventually stop. There are exceptions to this rule, such as a PLC program needing to continuously loop or a **user interface** (**UI**) needing to be constantly able to accept user inputs. Regardless of the type of loop, the loop will require some condition that is used to terminate the loop. As long as the expression is TRUE, a loop will continue to iterate over a block of code. If the condition fails to evaluate to FALSE, the loop will not terminate, and you'll have what is called an infinite loop.

Infinite loops are generally considered bugs and usually stem from a flawed logical expression in the termination section of the loop. Any type of loop can result in an infinite loop; however, they are mostly caused by WHILE loops and, in terms of ST, REPEAT loops. A FOR loop can also cause an infinite loop condition; however, it is much harder to cause an infinite loop with a FOR loop in ST.

An infinite loop can cause a plethora of problems, such as the program hanging up, consuming hardware resources, which can cause the system to lock up, and, in terms of PLCs, crashing the system altogether. Therefore, it is vitally important to have a logical expression that will terminate the loop.

Now that we've explored the basics of how loops work, let's look at designing them!

Exploring loops in pseudocode and flowcharts

The nature of loops can often confuse inexperienced programmers, especially when designing a program. Typically, the easiest way to hammer in the concept is to first look at the concept of loops in terms of both pseudocode and flowchart examples. Therefore, in this section, we're going to look at designing loops in both pseudocode and with a flowchart.

Exploring loops with pseudocode

Depending on what you're working on or where you're at in your programming journey, working out a loop with pseudocode can often greatly help conceptualize the needed logic. To teach the concept of loops, a good example to work through is a blinking light program. The steps to make a light blink are straightforward. The program will turn on a light, wait for a period of time, and then turn the light off. However, this will only flash the light once, and we want to continuously blink the light. To accomplish this, we need to loop the program. For this example, we can use the following pseudocode:

```
Top of loop:
    Turn on light
    Wait for 1 second
```

```
        Turn light off
        Wait for 1 second
  Go to top of loop
```

As soon as this program starts, the program will enter the loop. The logic in the loop will turn on, wait for one second, turn off, wait for one second, and once all that is done, it will start the process over again. This means that this particular program will cause the light to blink indefinitely.

The key here is the loop. In practice, all loops will have a starting point and an ending point. When the loop reaches the end of the loop, it will go back to the top of the loop. So, in this case, the loop will continuously run, and we'll have an infinite loop. Now, this isn't ideal. Typically, we only want to blink a light under certain conditions, such as an emergency or when a part run has been completed. For this program to serve a useful purpose in a PLC-based machine, we should use a precheck loop. So, let's modify this example to only blink a light when there is an emergency.

For this next design example, let's assume that when the temperature of an oven is over 200 °F, the PLC will blink an emergency light. However, when the temperature falls below 200 °F, the light will stop blinking. With these requirements, we can design our program with the following pseudocode:

```
Read temp
While temp > 200:
       Turn light on
Wait 1 second
Turn light off
Wait 1 second
End Loop
```

As can be deduced in this example, there is a conditional in the `While` line. This means that as long as that condition is TRUE, the blink code will run. If you recall, we want the light to blink only when our temperature reading is above 200. The condition in the example will only execute the blink code when the temperature is above 200. To summarize, when the temperature is above 200, the condition will result in a TRUE condition. When the temperature is below 200, the condition will result in a FALSE condition and the loop will break.

Why not use a counter loop or a post check loop for this program? The answer is simple. A counter or FOR loop will cause the light to blink a certain number of times. So, when the program reaches to loop, it would blink the light a certain number of times, no matter what. On the other hand, if we used a post check loop such as a REPEAT loop, the light would blink at least once, no matter what. In this case, we only want to blink the light when the temperature is above 200 and we want the light to blink indefinitely until the temperature is below 200; as such, the WHILE loop or precheck loop is the most appropriate for this task because we are not guaranteed that the code in the loop will run.

Now that we've seen a pseudocode loop, we need to see how these look visually. To do this, we're going to draw out the loop in a flowchart.

Representing a loop in a flowchart

Figure 13.1 shows what the blinking light program would look like as a flowchart:

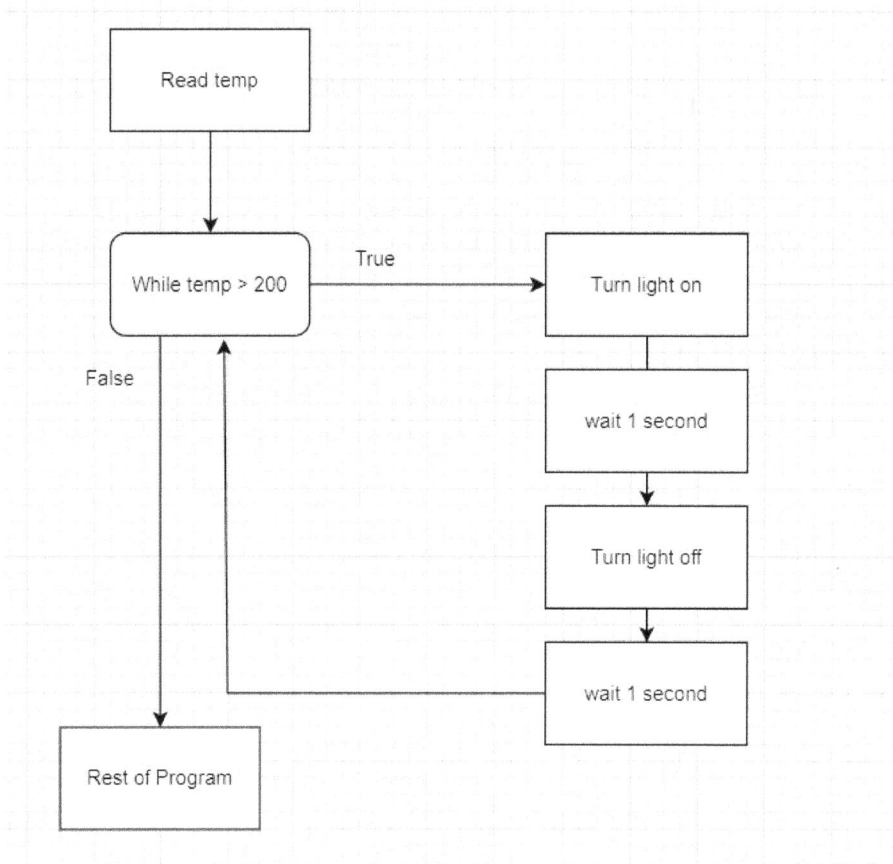

Figure 13.1 – The blinking light flowchart

In this diagram, we're using a rectangle with rounded edges to represent the loop. Many flowcharting systems will use a derivative of a rectangle or even the same diamond we use for an IF statement to represent a loop. To avoid confusion, in this chapter, we're going to use the rounded edge rectangle to represent the loop. However, keep in mind that different symbols may be used in different organizations and programs.

Regardless of what symbol you use, the core to understanding the program is following the arrows. If you follow the arrows, you will see that there is a loop with a logical expression in it. When the logical expression evaluates to TRUE, it begins the code sequence to blink the light. When the IF condition is FALSE, the loop will either terminate or be skipped altogether, and the program will continue on.

Not all loops have to be drafted out. For simple loops such as the blinking light program, developers will typically not draft out the program. However, it is wise to either use pseudocode or a flowchart for loops that are complex or contain a lot of logic. If you're new to programming, loops can be very confusing, and analyzing their execution can be quite difficult. Therefore, when you're first learning how to utilize loops, it is a good idea to use a flowchart or pseudocode to map out their execution.

Now that we've drafted out a few loop examples, we can move on and see loops in action. With that, let's explore the popular FOR loop!

Implementing a FOR loop

Since the FOR loop is a counter loop, it needs three things to function. The needed components are as follows:

- A starting value (typically 0, but it can be other values)
- A termination condition
- Incrementing/decrementing logic

In the IEC 61131-3 standard, there is a fourth, optional component that can be used to determine the step. In other words, it controls how much the loop will increase or decrease by. By default, the FOR loop will increase by 1. The general syntax for the FOR loop is as follows:

```
FOR <variable> := start_value to end_value <BY step> DO
    Code
END_FOR;
```

In this example, the BY step is the optional code that will increase or decrease the FOR loop by a stipulated value.

As with every other concept explored in this book, let's implement the loop to see how it behaves. For this example, let's create a loop that will count from 0 to 100. To do this, we will need to implement the following variables:

```
PROGRAM PLC_PRG
VAR
    counter : INT;
    iteration : INT := 0;
END_VAR
```

The counter variable will be responsible for keeping track of which iteration the loop is on. In other words, this variable will ultimately be responsible for the loop's starting point, termination, and incrementation. The iteration variable will serve as our dummy logic. For this program, all that variable will do is increase by 1 each time the loop iterates. This variable is not necessary for the loop to function; it only serves as something for the loop to do.

The logic for this example is as follows:

```
IF counter <= 0 THEN
    FOR counter := 0 TO 100 DO
        iteration := iteration + 1;
    END_FOR
END_IF
```

For this example, we have the FOR loop wrapped in an IF statement. In general, the IF statement is not necessary for the FOR loop to function. We are only using it in this example to control the loop's execution, so the FOR loop will terminate and not restart when the program loops back on itself. Again, it is important to remember that all PLC programs will automatically loop, so essentially, the FOR loop is a loop inside of a loop. If the IF statement is omitted, the loop will increment forever.

When the program is executed, you should be met with *Figure 13.2*:

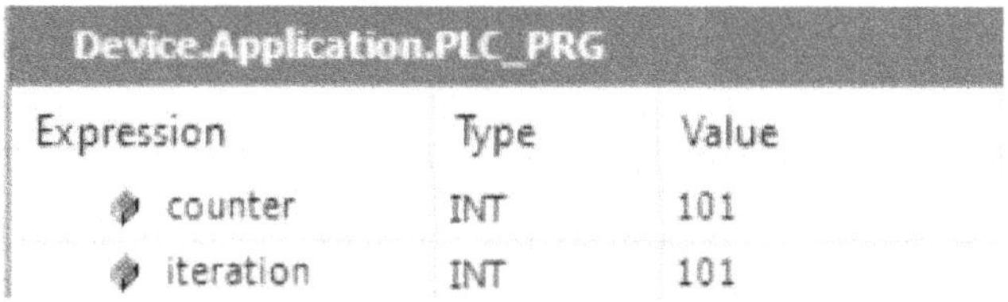

Device.Application.PLC_PRG		
Expression	Type	Value
◈ counter	INT	101
◈ iteration	INT	101

Figure 13.2 – A FOR loop execution

As you can see, both variables were initialized to. Now, consider the values in the variables. Our loop is programmed to stop at 100; however, the values both read 101. A logical question is "why"? The reason why we have 101 instead of 100 has to do with our initialization of 0 for both variables. Since our initial value in the FOR loop is 0, we end up with what seems an extra incrementation in our execution.

If you ran the example, you may have noticed that the iteration essentially jumped to the final value – in this case, 101. For demonstration purposes, we can slow down the counting by adding two TON function blocks and an extra start variable. To demonstrate this, we can add the following variables:

```
PROGRAM PLC_PRG
VAR
    start : BOOL := FALSE;
    t1 : TON;
    t2 : TON;
    counter : INT := 10;
    iteration : INT := 0;
END_VAR
```

The logic for this example will be as follows:

```
IF start = TRUE AND iteration <> 10 THEN
    FOR counter := 1 TO 10 DO
        t1(IN:= start AND t2.Q = FALSE, PT:=T#1S);
        t2(IN:=t1.Q, PT:=T#1S);

        IF t2.Q THEN
            iteration := iteration + 1;
        END_IF

    END_FOR
END_IF
```

For this example, ensure that iteration is set to a value that is not 10 and set start to TRUE. When this code runs, the iteration variable will increase about once a second until it reaches 10. When the variables reach 10, the program will halt its execution. For simplicity, we will not slow the loops that will be explored throughout the rest of the chapter. However, you can add the TON variables to the future examples and copy the code in the body of the FOR loop. From there, you can modify the code in the IF statement to count or do whatever else you want it to do!

Now that we have a basic understanding of the FOR loop, let's take a look at changing the step of the counter. To do this, simply modify the main logic to match the following code:

```
IF counter <= 0 THEN
    FOR counter := 0 TO 100 BY 2 DO
        iteration := iteration + 1;
    END_FOR
END_IF
```

The only difference between this FOR loop and the previous examples is the BY 2 code. With this additional code, the loop will increment by 2 instead of its default value of 1. To run this example, set the iteration and counter variables to 0. You're also going to want to remove any other variables that may have been implemented in the slow FOR loop example. When you're all set up, and run this example, you should be met with *Figure 13.3*:

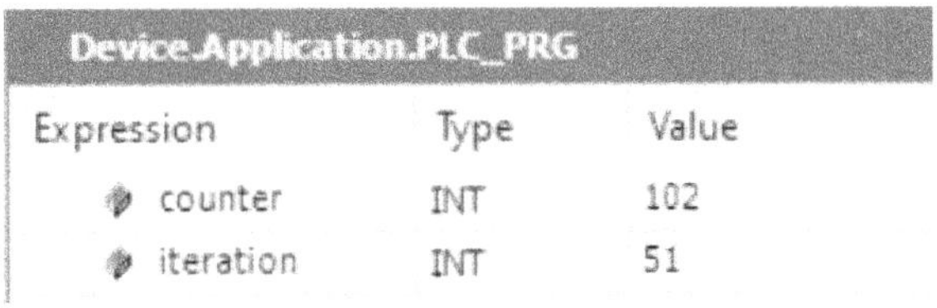

Device.Application.PLC_PRG		
Expression	Type	Value
counter	INT	102
iteration	INT	51

Figure 13.3 – A FOR loop with BY

In this output, you may notice that the iteration will only stop at 51 while the counter variable is 102. These values are due to the new step value. This means that we are only going to loop 51 times instead of 100 times because we are incrementing by 2.

All things considered, changing a step value is something that is either going to be done a lot on a project or never done at all. Typically, programmers will stick with the default incrementation of 1. However, knowing how to change the step is nonetheless important.

As stated before, the FOR loop is the most commonly used type of loop in programming but it's not the only one. The next step in understanding loops is learning how to implement a precheck loop. Therefore, the next section is going to be dedicated to implementing the WHILE loop.

Implementing the WHILE loop

The second most common type of loop is the almighty WHILE-DO loop. In everyday speech, most PLC developers will refer to the loop simply as a WHILE loop. The WHILE loop is a precheck loop, which means the first operation that happens is a check that will determine whether the loop should run at all. The general syntax for the WHILE loop is as follows:

```
WHILE <condition> DO
    Code
END_WHILE
```

For this syntax, the <condition> can be any logical expression. You can use the same logical expression or logical operators that you use with the IF statement in that expression.

To see this code block in action, let's look at an example. The first thing to do is set up the necessary variables. For this project, we're going to use the following:

```
PROGRAM PLC_PRG
VAR
    condition : INT;
    msg       : WSTRING;
END_VAR
```

The condition variable will hold a value that will determine whether the loop should start or continue, while the msg variable will hold a certain message while in the loop and another message when outside the loop.

The main logic for this example will be as follows:

```
WHILE condition >= 10 DO
    msg := "in loop";
END_WHILE
msg := "outside of loop";
```

This program will loop as long as the condition value is set to a value of 10 or greater. As long as the loop is running, the msg variable should say in loop. However, when the loop terminates or is skipped, it will continue to the last line of the program and change the loop to outside of loop.

To test the program, start it up and set the condition to a value of 10. Once you start the program and set the value, you should be met with *Figure 13.4*:

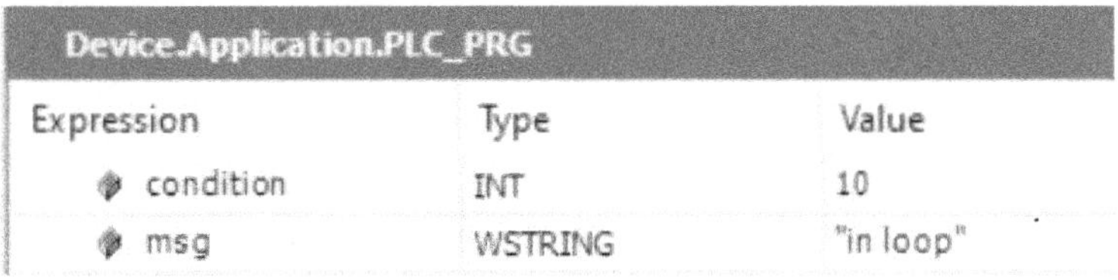

Device.Application.PLC_PRG		
Expression	Type	Value
♦ condition	INT	10
♦ msg	WSTRING	"in loop"

Figure 13.4 – In the WHILE loop

Now, set the conditional value to a value less than 10, such as 9. When you do that, you should see *Figure 13.5*:

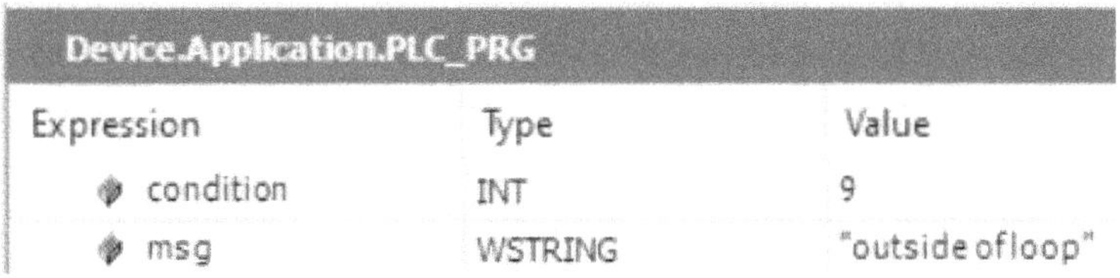

Device.Application.PLC_PRG		
Expression	Type	Value
♦ condition	INT	9
♦ msg	WSTRING	"outside of loop"

Figure 13.5 – Outside the WHILE loop

As can be seen, once the value is less than 10, the conditional in the WHILE loop will result in a FALSE state and the loop will terminate. When the loop terminates, the program will be allowed to move on and the msg will change.

It is important to remember that this is a precheck loop. This means that when you first start the program, this message may read that it is outside of the loop. This is accurate, as the default value for the variable is 0. This means that the loop will be skipped over when the program is first started. Therefore, if you see this when you first fire up the program, rest assured that everything is working as it should.

Now that we've explored counter loops (FOR loops) and precheck loops (WHILE loops), we can move on and explore post check loops.

Exploring the REPEAT loop

The REPEAT-UNTIL loop, often simply called a REPEAT loop, is very similar to the WHILE loop that we explored in the previous section. However, as we established before, the REPEAT loop is a post check loop, which means that we are guaranteed at least one execution of the code in the loop body.

The syntax for the REPEAT loop is as follows:

```
REPEAT
     Code
UNTIL <condition>
END_REPEAT
```

The conditional is at the bottom of the loop, which is why it is a post check loop. Just like with the WHILE loop, the condition can be any logical expression that we've seen thus far. As long as the condition evaluates to FALSE, the code in the loop will continue to run.

As with all the other loops that we've explored thus far, we're going to utilize an example to observe the loop in action. To do this, we're going to implement the following variables:

```
PROGRAM PLC_PRG
VAR
     testVar : INT := 10;
     inLoop  : BOOL := FALSE;
     msg     : WSTRING;
END_VAR
```

In this case, we have three variables. The first is testVar. This variable will simply serve as a variable that the loop can test against; by default, this variable will be initialized to 10. The next variable is the msg variable. This variable will hold a message to determine whether the program is in the loop or not. Lastly, the inLoop variable will be initialized to FALSE but will only toggle to TRUE when the code in the loop runs.

The main logic for the program should look like the following:

```
REPEAT
     msg := "less than 12";
     inLoop := TRUE;
UNTIL testVar > 12
END_REPEAT
msg := "greater than 12";
```

When the code is executed, you should be met with *Figure 13.6*:

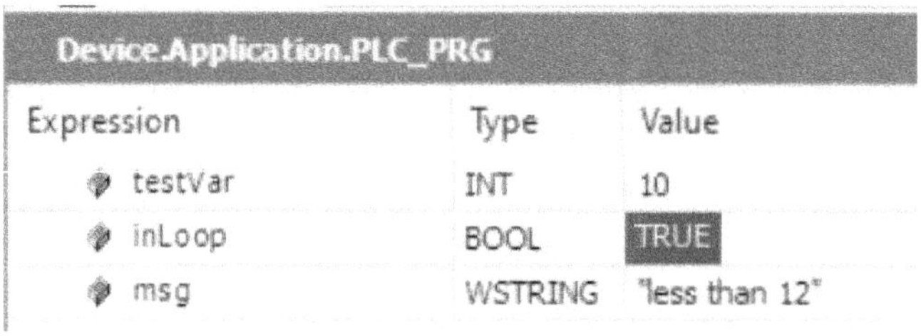

Device.Application.PLC_PRG		
Expression	Type	Value
testVar	INT	10
inLoop	BOOL	TRUE
msg	WSTRING	"less than 12"

Figure 13.6 – A False condition

Since 10 is not greater than 12, the conditional will result in a FALSE. As such, inLoop will toggle to TRUE and the message will say less than 12 because it is stuck in the loop. If you set the testVar variable to 20, the loop will terminate and change the message, as in *Figure 13.7*:

Device.Application.PLC_PRG		
Expression	Type	Value
testVar	INT	20
inLoop	BOOL	TRUE
msg	WSTRING	"greater than 12"

Figure 13.7 – The True condition

In this case, the message changed because we broke out of the loop. If you want to slow this example down, to observe it more closely, you can add the TON function blocks as well as the supporting code that was explored in the FOR loop sections.

Now, using conditionals is a great way to terminate a loop. However, there will come a time when a loop must be terminated when there is an unexpected circumstance that isn't supported in the loop's termination logic. To do this, the EXIT keyword can be used to prematurely exit the loop.

Exporting the EXIT keyword

The EXIT keyword will terminate a loop without the need for the loop's termination condition to run. For example, suppose you have a counter loop and you encounter a number that, for some reason, can cause an unsafe condition, and the loop needs to be immediately exited out of. Also, suppose that this value may not appear often during a program run. In cases such as this, the EXIT keyword can be used to terminate the loop before the main termination logic is executed. To demonstrate this, we're going to modify the FOR loop example from before. This example will have the same general structure as the past FOR loop example but will include a couple of extra variables:

```
PROGRAM PLC_PRG
VAR
    counter   : INT;
    iteration : INT := 0;
    loopPos   : WSTRING;
    msg       : WSTRING;
END_VAR
```

For this example, the counter and the iteration variables will serve the same purpose as they did in the past examples. However, this time, the program will include the msg variable and the loopPos variable. The msg variable will hold a message stating the iteration is equal to 5, while the loopPos variable will show where the program is in its execution.

The main logic for this program will be as follows:

```
IF counter <= 0 THEN
    FOR counter := 0 TO 100 DO
        iteration := iteration + 1;
        IF iteration = 5 THEN
            msg := "iteration = 5";
            EXIT;
        END_IF
    END_FOR
    loopPos := "outside of loop";
END_IF
```

For this example, we're setting up a FOR loop that should iterate from 0 to 100. However, the catch for this example is that when the iteration is equal to 5, we're going to set our msg variable to iteration = 5 and we're going to terminate the loop. We'll be able to tell whether the loop is terminated with the loopPos variable, as it'll set the message to outside of loop when the program exits out of the loop either by the EXIT command or by simply letting the loop terminate on its own.

When the program is executed, you should be met with *Figure 13.8*:

Device.Application.PLC_PRG		
Expression	Type	Value
◈ counter	INT	4
◈ iteration	INT	5
◈ loopPos	WSTRING	"outside of l...
◈ msg	WSTRING	"iteration = 5"

Figure 13.8 – The Exit command in action

As can be seen, *Figure 13.8* shows the loop terminating when the iteration is 5, as opposed to 100 as it did in the first example.

There is a gotcha with the EXIT command that many inexperienced programmers will stumble into. In short, it will be the last command executed in the loop. You can obviously have commands after the EXIT command, like in the example; however, no command that is in the execution line will run after the EXIT command. This means that the EXIT command will effectively kill the FOR loop until the program circles back around on it.

The EXIT command is mostly used to terminate a loop when there is a condition that invalidates the code. Suppose you have a program that depends on a calculation with division and for whatever reason, the denominator gets set to 0, which will result in a catastrophic program error. In a case such as this, the EXIT command will be embedded in either an IF statement, such as the one in the example, or some other type of error handler block.

Though we only demonstrated the EXIT command in a FOR loop, it can be used in any of the other types of loops. When it comes to a REPEAT or WHILE loop, it is best to try to use its own conditional expression to terminate the loop, as EXIT commands can add confusion to the loop's execution. However, the EXIT command will terminate the loop at the position it is called, so if certain logic needs to be ignored, you'll usually put in a control statement with the EXIT command before that potential erroneous area.

By this point, we've covered the basics of loops and how they work. We've explored how all the loops work and seen them in action. We've even learned how to prematurely terminate them. This means the next step is to learn how to embed them.

Understanding nested loops

Sometimes, we will need to iterate over another loop. To accomplish this, we use what are called **nested** or **embedded** loops. Essentially, a nested or embedded loop is a loop within a loop.

To conceptualize this, let's look at some pseudocode:

```
For counter1 = 1 to 100 do
     Print "counter 1 is:" + counter1
     For counter2 = 1 to 50
          Print "counter 2 is: " + counter2
     End_For
End_For
```

If you've never seen a nested loop before, the output might be hard to picture. To alleviate this, consider the following theoretical output:

```
counter 1 is: 1 <- Outer loop iterates
counter 2 is: 1
counter 2 is: 2
  .

  .

  .
counter 2 is: 50
counter 1 is: 2 <- Outer loop iterates
counter 2 is: 1
counter 2 is: 2
```

```
    .
    .
    .
counter 2 is: 50
counter 1 is: 3 <- Outer loop iterates
```

As you can see in the theoretical output, `counter1` will fire and `counter2` will loop 50 times, then `counter1` will iterate once more, causing the inner loop (`counter2`) to loop 50 more times. The flow for this program can be kind of hard to visualize for inexperienced programmers. For applications such as this, a flowchart is often handy to visualize the flow of the PLC program. As such, consider *Figure 13.9*, which is the graphical representation of the presented pseudocode:

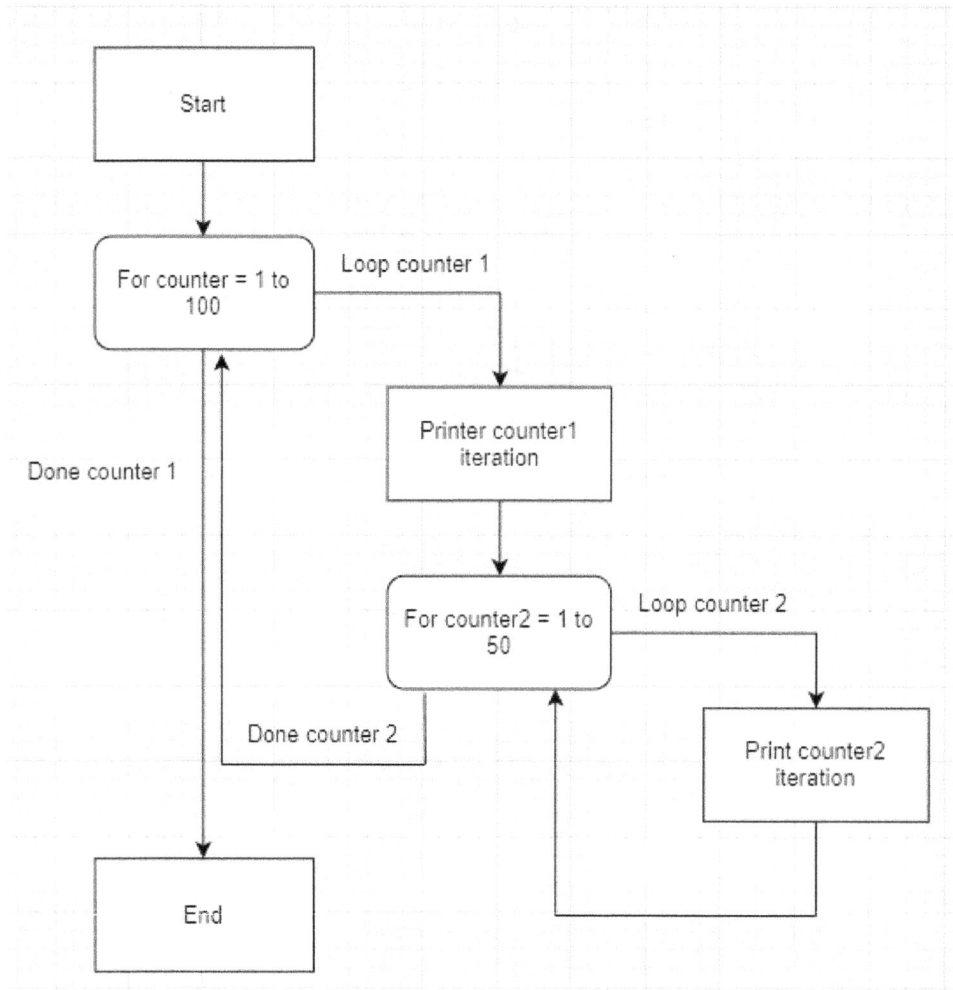

Figure 13.9 – The flow of a nested loop

Figure 13.9 may look confusing at first. However, to understand this flowchart, follow the arrows from top to bottom. If you look at the flow of the program, the program will start and then move to the first loop (`counter1`). The first loop will iterate 100 times, and with each iteration, the inner loop (`counter2`) will iterate 50 times. Once the first loop iterates 100 times, the program will end.

Challenge – creating a behavior report

If you followed along with the examples, you should have a decent grasp of how to implement different types of loops. A nested loop is just a loop within a loop. So, for a challenge, convert the program represented in the pseudocode and flowchart (both the same program) into a working PLC program.

Now, a common task that most PLC programmers will be forced to do on a regular basis is to write small memos or reports about a machine's behavior. Typically, these reports will inform laypeople or other engineers about problems, machine behaviors, and so on. These reports will usually involve you, as the engineer/programmer, making a series of observations about a machine and putting those observations into layperson's terms. So, for practice, make the following observations and write a short report of around 250 words that addresses the following questions:

1. How many times does the `counter2` loop iterate in total?
2. How many times does the `counter1` loop iterate in total?
3. What is the final output of the program?
4. Does the final output of the program meet expectations?
5. What can you do to ensure the program does meet expectations?

Also, note that if you simply translated the code without any control statements, the program would simply count to infinity due to the looping nature of a PLC program.

Take some time to write this report. Believe it or not, one of the most valuable skills a PLC developer can have is the ability to effectively convey information via text.

When you complete this challenge, you can move on to the final project.

Final project

For this project, we're going to make a hypothetical factory that has three assembly line states. As stated in the chapter introduction, one assembly line state will create a certain number of parts, another will create at least one part, and finally, the last line state will only create parts under certain conditions. To do this, we're going to need a simple `CASE` statement to direct the part down the correct assembly line.

In terms of real-world operations, a setup such as this isn't unheard of. Depending on the type of assembly/process, a machine may be put into certain modes either to produce a certain number of parts, examine previous process results to determine whether the part should continue production, or create a test product for quality assurance or whatever other reasons. Typically, applications such as these are controlled with state machines. To begin, let's create a pseudocode mockup.

Design

This program is going to be a simple CASE statement. When the assembly line number is selected, the program will go into that CASE statement. For example, if the user selects 1, the program will create a certain number of parts, while an input of 2 will create at least one part, 3 will not guarantee a part to be created, and finally, 4 will put the machine into a standby mode:

```
Case assemblyLine
1:
      FOR numOfParts := 1 to parts
          Create Part
      END_FOR
2:
      REPEAT
          Create Part
      UNTIL partCreated >= numOfParts
      END_REPEAT
3:
      WHILE partsCreated < numOfParts DO
          Create Part
      END_WHILE
4:
      Put machine into standby
```

Now that we have a design in place, we can move on to implementing the working program.

Code implementation

With a solid design in place, let's start implementing the code. As usual, we're going to start with creating the variables for the program:

```
PROGRAM PLC_PRG
VAR
      assemblyLine : INT;
      numOfParts   : INT;
      parts        : INT;
      partsCreated : INT := 0;
END_VAR
```

For this project, there will only be four integer variables. The `assemblyLine` variable will be used to determine which line will be used, the `numOfParts` variable will be used to determine the number of parts that will be created, and the `parts` variable will be used to hold the number of parts that need to be created. The last variable, `PartsCreated`, will be initialized to 0. This variable will keep track of the number of parts that have already been created.

Once the variables are created, we can implement the main logic:

```
CASE assemblyLine OF
1:
      partsCreated := 0;
      FOR numOfParts := 1 TO parts DO
            partsCreated := partsCreated + 1;
      END_FOR
2:
      partsCreated := 0;
      REPEAT
            partsCreated := partsCreated + 1;
      UNTIL partsCreated >= numOfParts
      END_REPEAT
3:
      partsCreated :=0;
      WHILE partsCreated < numOfparts DO
            partsCreated := partsCreated + 1;
      END_WHILE
4:
      numOfParts    := 0;
      parts         := 0;
      partsCreated := 0;
END_CASE
```

This implementation is pretty straightforward and closely follows the pseudocode with some minor tweaks. For example, we set `numOfParts` to 0 in each case. We do this to reset that value for a fresh run, as the initialization value will be overwritten after the first run. Technically, you don't have to initialize that value, but it is typically a good idea just as a sanity check. We also defined what *case 4* actually is in terms of functionality – that is, it sets all the values to 0.

Once you've finished implementing the logic, you can move on to testing it.

Testing the program

In these sections, we are going to be looking at testing each case. Since there are a few different cases to check, these sections will be broken down into a series of subsections that will cover each case. As such, if you are more comfortable with a certain loop, you can start at that section and work your way up to harder ones.

Case 1 – creating a certain number of parts with a FOR loop

The first simulated assembly line state that we're going to test is the one that will always make a certain number of parts, *case 1*. A real-world example of this line state would be a mode that creates a certain number of specialized parts. To run this line, we will need to choose this case or state as well as provide the number of parts we want to make. For this example, we're going to create five parts in total. So, for this example, write 1 to `assemblyLine` and 5 to `parts`:

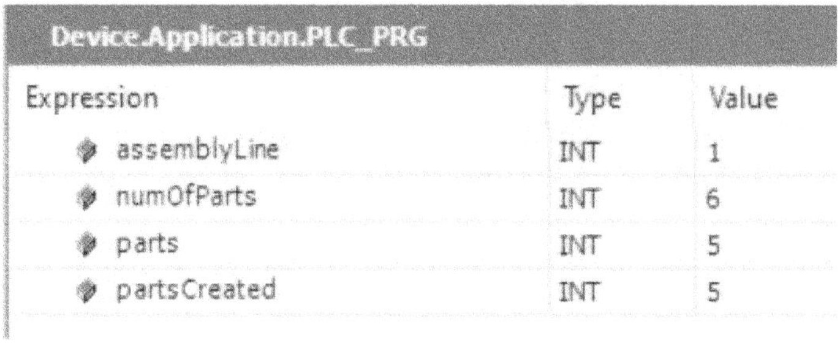

Device.Application.PLC_PRG		
Expression	Type	Value
assemblyLine	INT	1
numOfParts	INT	6
parts	INT	5
partsCreated	INT	5

Figure 13.10 – The FOR loop assembly line

The number to watch is the `partsCreated` variable in the bottom row. Notice that we wanted to make five parts. Also, notice that the value in that row is 5. As such, we can call this assembly line a success, since it essentially says it made five parts!

With this line tested, let's move on to the second line.

Case 2 – making at least one part with a REPEAT loop

This line state is going to be a little tricky to test. We're always going to make one part, no matter what. In essence, what we're doing is creating a series of test parts. So, the first thing we're going to do is reset all the variables back to 0 and simply write 2 to the `assemblyLine` variable. When you do this, you should be met with *Figure 13.11*:

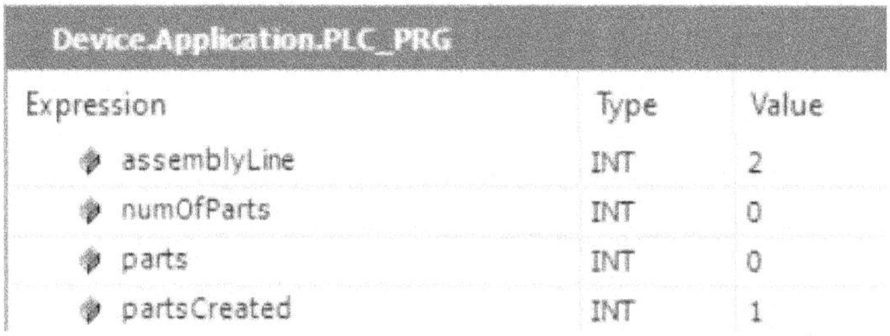

Device.Application.PLC_PRG		
Expression	Type	Value
assemblyLine	INT	2
numOfParts	INT	0
parts	INT	0
partsCreated	INT	1

Figure 13.11 – Producing one part by default

Now, if you notice the last row in the screenshot, you will see that the number of parts we created was 1. This is because we are using a REPEAT loop, and we are guaranteed at least one iteration, or, in this case, one part. So, we can safely say that this logic is working.

Now, we need to move on and test it to ensure it is producing the correct number of parts. To do this, write 5 to numOfParts:

Device.Application.PLC_PRG		
Expression	Type	Value
◈ assemblyLine	INT	2
◈ numOfParts	INT	5
◈ parts	INT	0
◈ partsCreated	INT	5

Figure 13.12 – Producing five parts

Again, if you examine the last row in *Figure 13.12*, you will see that the number of parts created is indeed 5. This means that, coupled with the other data from the first test, the state is working as intended, and we can now move on to test the final assembly line state!

Case 3 – creating a specified number of parts with a WHILE loop

The last assembly line we're going to test is the one that will continuously create parts. In a real-life setting, this line would be a main production line that would make parts until a condition, such as pressing a button, is performed by an operator. In this example, we have the functional equivalent of a FOR loop. The reason for this is to avoid a runaway loop that would produce different results between the test data presented here and the value that you, the reader, may get when you run your example. So, just be warned that in a real-life situation, you wouldn't use logic like in the example; you would really use a Boolean variable and have the process repeat until that variable is FALSE!

With that out of the way, our goal will be, again, to make five parts. However, since this is a production line, we need to ensure that when no parts are ordered, the assembly line is virtually off. To do this, set all the variables to 0, with the exception of the assemblyLine variable, which should be set to 3:

Device.Application.PLC_PRG		
Expression	Type	Value
◈ assemblyLine	INT	3
◈ numOfParts	INT	0
◈ parts	INT	0
◈ partsCreated	INT	0

Figure 13.13 – Do not create any parts

In the case of *Figure 13.13*, our goal was to not create any parts. In this case, we used a WHILE loop, which means that we are not guaranteed any parts. If you think about the logic, you will see that this line should only run when the number of parts created (partsCreated) is less than the number of parts ordered (numOfParts). This means that the data in *Figure 13.13* is indeed as it should be, and this test case worked!

So, now that we know our line will not produce any parts until we essentially tell it to, let's create five parts. To do this, simply write 5 to numOfParts and you should be met with *Figure 13.14*:

Device.Application.PLC_PRG		
Expression	Type	Value
⊕ assemblyLine	INT	3
⊕ numOfParts	INT	5
⊕ parts	INT	0
⊕ partsCreated	INT	5

Figure 13.14 – Creating five parts on line 3

Now, as can be seen, we have five parts created, so this line works. This means we can now move on to the machine standby state.

Case 4 – machine standby

This case is very simple. All it does is reset all the variables to 0. In other words, if this were a real machine, this case would reset all the data and wait for further instructions. So, to test this case, all you have to do is write 4 to assemblyLine and you should be met with *Figure 13.15*:

Device.Application.PLC_PRG		
Expression	Type	Value
⊕ assemblyLine	INT	4
⊕ numOfParts	INT	0
⊕ parts	INT	0
⊕ partsCreated	INT	0

Figure 13.15 – Machine standby

If all goes well, your output should match that of *Figure 13.15*. If it does, this means you have a working factory!

Summary

In this chapter, we explored a variety of loops such as the FOR loop, WHILE loop, and REPEAT loop. We also explored the theoretical side of loops, such as what counter, post check, and precheck loops are, as well as all the mechanics that go behind them. In all, you should have a decent understanding of how loops work by this point. Now, if you are still a little unsure, that's okay. Loops can be tricky to understand, and many entry-level programmers will often struggle with them at first. If you do find yourself in this category, please go back and work through the example again and try to make a few of your own. Above all else, be patient and practice!

Now, there are many applications for loops. What we explored in this chapter barely scratched the surface of what loops are capable of and how they are used in a real-world setting. In the next chapter, we are going to explore how loops can help us do a very vital task: sorting!

Questions

1. What type of loop is a WHILE loop?

2. What type of loop is a FOR loop?

3. What type of loop is a REPEAT loop?

4. What is a post check loop?

5. What is a precheck loop?

6. What is a counter loop?

7. How many iterations are you guaranteed with a WHILE loop?

8. How many iterations are you guaranteed with a REPEAT loop?

9. When will a FOR loop terminate?

10. What does the EXIT command do?

11. Does the FOR loop use a custom logic statement to terminate?

    - Yes

    - No

Further reading

- *FOR Statement*: `https://www.fernhillsoftware.com/help/iec-61131/structured-text/st-for.html`

- *WHILE Statement*: `https://www.fernhillsoftware.com/help/iec-61131/structured-text/st-while.html`

- *REPEAT Statement*: `https://www.fernhillsoftware.com/help/iec-61131/structured-text/st-repeat.html`

Part 3:
Algorithms, AI, Security, and More

This part will introduce exotic concepts and emerging technologies that are usually glossed over in automation programming. This part will be an applied section that is built on previous chapters. It will cover concepts such as generative AI (ChatGPT), provide basic cybersecurity awareness for PLC-based systems, explore sorting algorithms, and more. The part will end with a comprehensive project that will cover topics that were covered throughout the book.

This part has the following chapters:

- *Chapter 14, Sorting with Loops*
- *Chapter 15, Secure PLC Programming – Stopping Cyberthreats*
- *Chapter 16, Troubleshooting PLCs – Fixing Issues*
- *Chapter 17, Leveraging Artificial Intelligence (AI)*
- *Chapter 18, The Final Project – Programming a Simulated Robot*

14

Sorting with Loops

In automation, it is often necessary to have to figure out the range of certain characteristics of the part your machine produces. For example, it is quite common to need to know how heavy the heaviest part was or how short the shortest part was. Sorting characteristics of parts is often required in many manufacturing environments to ensure proper quality. Typically, to effectively perform the necessary statistics calculations, it is helpful to have all the data sorted.

Sorting is a vital concept in computer science. Sorting can be a very costly operation in terms of memory, CPU power, and most importantly, time. As logic dictates, the more data you have to sort through, the more time and resources you will need at your disposal. Therefore, when sorting, you have to choose an appropriate sorting methodology that is easy to implement while still being efficient enough to get the job done in a timely manner.

The key to sorting is looping. More specifically, FOR loops are a vital component of sorting. This chapter is going to be dedicated to learning about sorting, sorting algorithms, and all the necessary components needed to efficiently sort. Now, before we proceed, this chapter is meant only to expose you to the basics of sorting and programming algorithms. Sorting is an advanced concept and requires advanced knowledge of pointers and scalable arrays, which goes beyond the scope of this book. However, after reading this chapter, you should understand the basics of sorting, algorithms, and the mechanics of sorting. To help you do this, we're going to explore the following concepts:

- What sorting is and why it is important
- The basics of arrays
- What is a sorting algorithm?
- Efficiency metrics with Big O and Big Omega
- Common sorting algorithms

Finally, to round out the chapter, we're going to implement one of the simplest yet one of the least efficient sorting algorithms there is, bubble sort, to find the heaviest and lightest bag of cement that was filled during a production run.

Technical requirements

The code for this chapter can be found at the following URL: `https://github.com/PacktPublishing/PLCs-for-Beginners`

Unlike with most chapters, you can pull down and explore the code, but the guts of the examples cannot be heavily modified. This may seem odd, but we'll dig deeper into this in the upcoming sections.

How to use this chapter

This chapter is going to focus on sorting and sorting algorithms. Sorting is one of many applications for algorithms. Sorting is a very important concept but as you work through this chapter, think of sorting as a catalyst for learning algorithms.

As we will see in the upcoming section, there are countless sorting algorithms and new algorithms are being invented all the time. This means that not every single sorting algorithm will be covered in this book. Learning algorithms is less about memorizing the actual algorithm and more about learning the mechanics of algorithms in general. So, as you work through this chapter, try not to focus on memorizing the algorithms presented and focus more on the core theory presented. Therefore, let's start off with learning what sorting is.

What is sorting?

Before you can implement an algorithm, it is very important that you understand what the core problem is that the algorithm is meant to solve. In other words, before you implement a sorting algorithm, you need to understand what sorting is. We all know that sorting is the systematic process of placing items in an increasing or decreasing order based on a certain characteristic. For example, in manufacturing, it is common to sort items such as bags of sand or cement in a certain order based on their weight or another characteristic, such as their overall size.

Sorting is a simple concept that can be done in many ways. This means that there are many different algorithms that can be used to sort items. In terms of computer science, sorting can be a very resource-intensive task. For example, a computer or PLC may need to sort through millions of data points to find the smallest or largest characteristic of a part or parts. Sorting through copious amounts of data points can tax the CPU and memory of a PLC to the max and, of course, take a prolonged amount of time. As such, as alluded to before, sorting has to be done correctly to ensure that the code is not only easy to maintain but also gets the job done quickly and efficiently.

So, how is sorting done? What are the needed components that a PLC programmer will need to implement an efficient sorting program? At a very high level, a programmer will typically use what is called an **array** and a **sorting algorithm** to sort items. As such, the first thing that we're going to explore is arrays. Arrays can be very complex and there is a lot that goes into using an array; however, the next section is going to be dedicated to the basics of arrays and how to use them. If you are not familiar with arrays, you must read the following section and understand it! A solid understanding of arrays is vital to sorting!

Exploring what arrays are and how to use them!

There are many times in programming when many related values will need to be sorted. For example, suppose you are creating a machine that can produce 100 parts for a given run. Now suppose that for each part, the machine will need to store the part's weight, length, and height. If you created a variable for each part's attribute, you would need to create 300 variables. Put simply, that would be a poorly implemented program that would be nearly impossible to troubleshoot and debug. A much easier approach would be to make three variables that can each hold 100 values. Until now, we haven't been able to do this. As we've seen until now, one variable holds one and only one value. So, how can we store 100 values in a single variable? Enter the world of arrays!

What is an array?

Put simply, an array is a variable that can hold multiple related values. The easiest way to think about an array is as the crew of a battleship. Just as each battleship has a crew that is made of multiple sailors who all have multiple jobs, an array is a set of multiple values that describe something such as a production run. The syntax for declaring an array in IEC 61131-3 is as follows:

```
Array_Name : ARRAY [starting point..end point] OF <type>;
```

If you've programmed in traditional languages such as C++ or Java, you may remember that arrays in those languages are 0-indexed. That means that the first element is 0. However, in IEC 61131-3, you can pick the first element. Therefore, it is not uncommon for a PLC programmer to see an array that starts at 1. Now, the endpoint is essentially the last value in the array. So, the whole syntax will determine the number the array starts with and the endpoint it ends at:

```
Array_Name : ARRAY [1..3] OF INT;
```

The array would essentially be `Array_Name[1]`, `Array_Name[2]`, and `Array_Name[3]`.

Now that we understand the array structure, we need to understand what array elements are.

Array elements

Array elements are essentially the individual variables in the array. In other words, you can think of `Array_Name[n]` as an element. This means that each `Array_Name` is an element and the `1..3` of the code snippet allocates the elements. Now, for an array to be useful, you need to be able to access each individual element in the array to retrieve the value that lives in it. To do this, consider the following scenario.

Suppose you have an array that was declared with the following snippet:

```
Array_Name : Array [0..10] OF INT;
```

This array starts at 0, or in regular programming lingo, is 0 indexed. Now assume for this example that we want to access the fourth element of the array. To do this, we would simply use the following code snippet:

```
Array_Name[3];
```

Notice that we used 3 instead of 4 to access the fourth array. This again stems from the array being 0 indexed. Now, this example is merely conceptual; in reality, these arrays are empty, which means that the array does not contain any values. As usual, to explore the behavior of an array, let's implement one in CODESYS.

Initializing an array

There are multiple ways to initialize an array. For this example, we're going to use a FOR loop to load data into the array. As such, the first thing we're going to do is declare our variables with the following code:

```
PROGRAM PLC_PRG
VAR
      sizes    : ARRAY [1..3] OF INT;
      i        : INT;
END_VAR
```

For this example, we're going to have an array called sizes that will have a total of three elements, starting at 1 and ending at 3. The i variable will be used as a counter variable for the FOR loop.

To load data into the array, we're going to use the following logic:

```
FOR i := 1 TO 3 DO
    sizes[i] := i;
END_FOR
```

When the program is executed, you should be met with *Figure 14.1*:

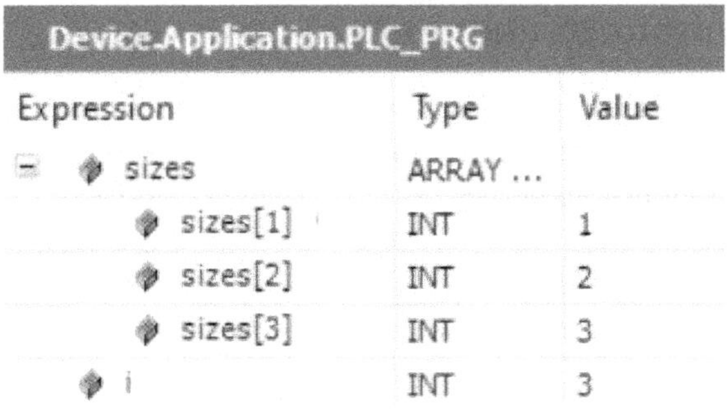

Figure 14.1 – The array output

Now, this is just one way of loading data into an array. Another way of loading data into the array is when you declare it. For example, you could use the following syntax to load five values into the array:

```
sizes : ARRAY [0..5] OF INT := [5,9,1,3,2,7];
```

When this code is executed, you should see the output in *Figure 14.2*:

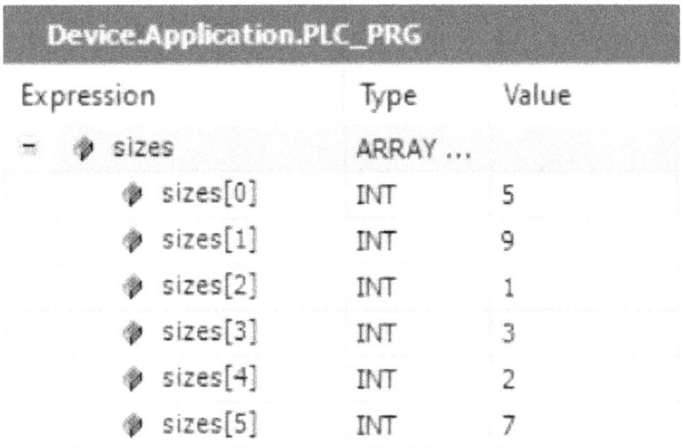

Figure 14.2 – An initialized array

For this book, we're going to use the second method to load data into an array. Arrays can be a bit confusing and can be considered a somewhat advanced concept to most. As such, we're going to use the second array method exclusively throughout this book.

Retrieving the number of elements in an array

For many applications, we will need to determine the size of the array. Depending on what you're doing, you may be working with multiple arrays that all have a different number of elements in them. This means that it is important to understand how to calculate the size of the array with code. To do this, we will use the SIZEOF command. In short, the SIZEOF command will return the size of something. Now, the size that the command returns is the size of an array in bytes. This means that to calculate the number of elements in the array, we must do a little math.

To calculate the size of an array, you would use the following code:

```
arrSize := SIZEOF(arry)/SIZEOF(arry[1]);
```

Essentially, this code will find the total bytes of the array and divide the number by the size of a single element. Since all the elements will be of the same size, the result will be the number of elements in the array! Consider the following example.

Suppose we have the following variables:

```
PROGRAM PLC_PRG
VAR
     sizes : ARRAY [0..5] OF INT := [1,9,5,3,2,7];
     arrSize : INT;
END_VAR
```

Put simply, we have an array and a variable called arrSize that will house the number of elements in the array.

In terms of logic, we will use the following:

```
arrSize := SIZEOF(sizes)/SIZEOF(sizes[1]);
```

For this example, we can use any element we want in the denominator of the equation. We're simply using element 1 because we know it should always be there. In all, you can use whichever element you want, but it is best to use element 1 since arrays typically start at either 0 or 1. When the code is run, we should get the following output:

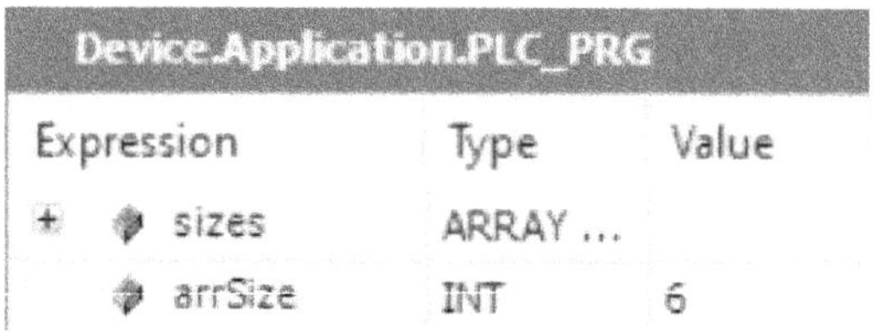

Device.Application.PLC_PRG		
Expression	Type	Value
+ ◆ sizes	ARRAY ...	
◆ arrSize	INT	6

Figure 14.3 – Array size

Now, this example can be thought of as a code pattern or code recipe. You can use this snippet any time you need to calculate the number of elements in an array. Typically, it is advisable to do this calculation anytime you are working with a sorting algorithm.

Now, a logical question is –"*Why are we exploring arrays in a chapter about sorting?*" Many students will often get frustrated with arrays when trying to master sorting algorithms. For an in-depth answer to why arrays are important, please read the next sections!

Why are arrays important for sorting algorithms?

Okay, now that we know how arrays work, why should we care about them, especially since we're learning about algorithms and sorting? Well, that's a logical question, and the answer is that most sorting and searching algorithms use arrays as a data structure to hold the data. In other words, you can think of arrays as holders to house the data for the given operation, whether that be sorting, searching, or whatever it may be. In all, once you get the array to output the correct data – that is, the data in *Figure 14.2* – you can move on and explore sorting algorithms!

Exploring sorting algorithms

What is a sorting algorithm? Better yet, what is an algorithm? To understand what a sorting algorithm is, one must first understand what an algorithm is in general. Therefore, the next section is going to be dedicated to understanding what an algorithm is at a high level!

What is an algorithm?

Often, computer scientists will casually throw the word "algorithm" around to represent any program. This is especially true in academic research. Now, technically, any program can be considered an algorithm, but the way the word is used by everyday programmers is to represent a specific series of instructions to accomplish a task.

There are many algorithms that do many different things. For example, if one were to look at machine learning, there are many different algorithms that can be used to implement neural networks, natural language processing, and many other advanced artificial intelligence attributes. On the other hand, there are also algorithms that perform simple tasks such as efficiently finding a single value in a large collection of values. A partial list of some common algorithm applications is as follows:

- Machine learning/artificial intelligence
- Hashing (security)
- Encryption/decryption (security)
- Searching (finding a value in a dataset)
- Sorting (sorting a dataset)

These are just a few areas where algorithms are used. Put bluntly, every aspect of computer science has algorithms associated with it. Regardless of the task, there is probably a set of instructions that can accomplish it. So, what exactly is a sorting algorithm?

What is a sorting algorithm?

A sorting algorithm is a specific algorithm that is designed to efficiently sort values in an array from greatest to largest or vice versa. As stated before, some sort of sorting algorithm is often employed during production runs to sort parts by certain attributes such as length, weight, height, and so on. Many times, the values are displayed on an HMI or other display unit so operators or engineers can easily see the upper and lower bounds of a production run.

Now, there are many different sorting algorithms that can be used. For example, one could opt to use the bubble sort, the heap sort, the merge sort, or any of the countless other sorting algorithms to accomplish their task. At the end of the day, each one of these algorithms will sort the values in an array; however, how efficiently they do so and how complex the code is will vary. To get a better grasp on efficiency, we're going to explore some efficiency metrics called Big O and Big Omega!

Algorithm efficiency metrics

It is important to understand how well your algorithm is going to perform. To understand this vital statistic, **Big O** and **Big Omega (Big Ω)** metrics are used. This section is going to be dedicated to exploring and understanding these metrics at a high level. Therefore, let's start our discussion with Big O!

Exploring the Big O notation

The most common efficiency metric for a sorting algorithm is the Big O notation. The Big O notation, or simply Big O, represents the upper bound of an algorithm's time complexity. In other words, in terms of sorting, you can think of Big O as the worst-case execution time for an algorithm. In terms of software development, you want as small a Big O value as possible. The following are some common Big O time complexities:

- **Constant time complexity**: Represented as $O(1)$, this is the most ideal time complexity. With this Big O time complexity, that worst-case runtime will never change, regardless of the number of elements in an array.

- **Logarithmic time complexity**: This is represented as $O(\log n)$, where n can be thought of as the number of items to sort or the number of elements in an array. This time complexity grows logarithmically with the number of items the algorithm has to sort.

- **Linear time complexity**: This is represented as $O(n)$, where n is the number of items to sort. This time complexity will increase linearly with the number of items.

- **Linearithmic time complexity**: This is represented as (n log n) where *n* is again the number of items the algorithms will sort. This is typically considered a good time complexity as many common sorting algorithms, such as merge sort and quicksort, utilize this time complexity. If you opt to use a prebuilt sorting algorithm, like one that found in a library, it will probably be one of those algorithms and have this particular time complexity.

- **Quadratic time complexity**: This is represented as O (n^2). This time complexity is most associated with bubble sort. This time complexity is typically not ideal, and depending on what you're doing, it is better to try to use a different algorithm.

There are many more common time complexities, and a very well-versed computer scientist can derive time complexities based on the algorithm; however, those concepts are beyond the scope of this book. In all, these are arguably the most common ones that a developer will run across in the field. Typically, a good developer will use an algorithm based on the worst-case scenario. Generally, no matter what you're doing, you want to develop your software for the worst conditions possible.

However, there is another metric that, though not as commonly used, is also very important to understand. That metric is called Big Ω.

Exploring the Big Ω notation

Sometimes, it is not enough to simply understand the worst-case scenario. Often, it is helpful to understand what the best-case scenario is for an algorithm. To accomplish this, the Big Ω notation is often used. In short, where Big O is the worst-case scenario, the Big Ω is the best-case execution for an algorithm. As can be assumed, the same notation that was explored before can also be used with Big Ω.

As stated before, the most common time complexity notation to use is the Big O notation. When you're asked about efficiency, it is usually assumed that the answer will be in terms of Big O. As such, though Big Ω is an important time complexity notation, this book will focus on implementing algorithms in terms of Big O.

With that, let's look at some common sorting algorithms!

Common sorting algorithms

There are more sorting algorithms than you can shake a stick at. Choosing the correct algorithm for a particular task will depend on factors such as the size of the dataset, how unsorted the dataset is expected to be, and so on. The following section is going to explore two of the most popular sorting algorithms and their possible use cases. The dataset we're going to use for this section is going to be the following array:

```
[1,9,5,3,2,7];
```

For this chapter, we're not going to screenshot the output for each algorithm. The way you can tell whether the algorithm worked is whether the data is sorted from least to greatest. To begin the discussion, we're going to look at the famous, or as some would call it, infamous bubble sort algorithm.

Exploring bubble sort

In academia, a student's first shake with a sorting algorithm is usually what's called the bubble sort algorithm. The bubble sort algorithm is typically used for small datasets or datasets that are nearly sorted. In terms of automation, that means arrays with small numbers of items in them or arrays where only a few numbers need to be swapped with numbers that are relatively close in the array. What the bubble sort lacks in speed, it makes up for in simplicity.

To implement the algorithm, first, start by implementing the following variables:

```
PROGRAM PLC_PRG
VAR
    sizes : ARRAY [1..6] OF INT := [1,9,5,3,2,7];
    i, j, temp: INT;
END_VAR
```

The size array is the set of values that will be sorted. The i and j variables are used as counter variables for the two loops in the algorithm, while the temp variable is used to swap values in the array. The main logic for the algorithm will look like the following:

```
FOR i := 1 TO 6 DO
    FOR j := 1 TO 6 - i DO
        IF sizes[j] > sizes[j + 1] THEN
            // Swap elements
            temp := sizes[j];
            sizes[j] := sizes[j + 1];
            sizes[j + 1] := temp;
        END_IF
    END_FOR
END_FOR
```

Bubble sort works by comparing numbers. If the value next to the value the algorithm is analyzing is larger, the algorithm will swap the two values. Hence, the temp variable. This process will continue until the values are all sorted. In other words, the larger values will bubble up the array.

Now that we've explored bubble sort, another simple algorithm to explore is the insertion sort algorithm!

Exploring insertion sort

Insertion sort is another $O(n^2)$ algorithm. This means that in terms of worst-case scenarios, it will perform in part with bubble sort. The algorithm works in a similar way to sorting cards in your hands. We start off with the first card, which we assume is sorted, then when we draw a second card, if it is greater, we put it to the right, or else we put it to the left. To see this algorithm in action, implement the following variables:

```
PROGRAM PLC_PRG
VAR
    sizes : ARRAY [0..5] OF INT := [1,9,5,3,2,7];
    i, j, key, temp: INT;
END_VAR
```

In this case, all we have is the array and a few variables that are going to be used in the main logic, which is as follows:

```
(* Insertion Sort Algorithm *)
FOR i := 1 TO 5 DO
    key := sizes[i];
    j := i - 1;

    WHILE j >= 0 AND sizes[j] > key DO
        sizes[j + 1] := sizes[j];
        j := j - 1;
    END_WHILE;

    sizes[j + 1] := key;
END_FOR;
```

Now, since we assume the first element is sorted, we start at 1 as opposed to 0. Outside of that, the algorithm is straightforward. For both these algorithms, you will be able to tell whether they are working by simply examining the output. In short, the values should be sorted from least to greatest.

Now that we have a rough idea of how to implement an algorithm, we can move on to a challenge.

Challenge – Merge sort

A more efficient sorting algorithm is the merge sort algorithm. The Merge sort algorithm has a Big O of $n*log(n)$. This algorithm is much more complicated to implement than bubble sort. However, it is one of the most used sorting algorithms. This algorithm uses what's called **divide-and-conquer**. In its most rudimentary sense, the algorithms will divide the problem (sorting array) into small portions, sort them, and then combine them again. Typically, this algorithm is divided up into multiple files called functions, or uses a data structure called a function block, which is like a class in C++ or Java.

The pseudocode for merge sort can be viewed in the following snippet. Typically, this algorithm uses a concept called **functions and recursion** to easily implement. However, since those are advanced topics, they are beyond the scope of this book. Therefore, a simplified version of the algorithm is presented further ahead.

Now, algorithms are generally presented in a language-agnostic manner. In other words, when you're presented with an algorithm, you are typically given pseudocode to translate into whatever language you're working with. This challenge is going to be a little more complex than the ones presented thus far in the book. For this challenge, take the following pseudocode example and convert it to a working program. This challenge will take some time, so be patient and revisit it if necessary:

```
IterativeMergeSort(arr):
  n = length(arr)
  for current_size = 1 to n - 1 by 2 * current_size:
      for left_start = 0 to n - 1 by 2 * current_size:
          mid = min(left_start + current_size - 1, n - 1)
          right_end = min(left_start + 2 * current_size - 1, n - 1)

          left_size = mid - left_start + 1
          right_size = right_end - mid

          left = arr[left_start to mid]
          right = arr[mid + 1 to right_end]

          i = 0
          j = 0
          k = left_start

          while i < left_size and j < right_size:
              if left[i] <= right[j]:
                  arr[k] = left[i]
                  i = i + 1
              else:
                  arr[k] = right[j]
                  j = j + 1
              k = k + 1

          while i < left_size:
              arr[k] = left[i]
              i = i + 1
              k = k + 1
```

```
while j < right_size:
    arr[k] = right[j]
    j = j + 1
    k = k + 1
```

Again, this challenge will take you some time to get working. However, you will need to practice this skill. It is a very common interview technique to have a candidate convert an algorithm in pseudocode to a working program. As such, getting as much practice as possible will only help you in the long run! Nonetheless, whenever you're ready to continue, you can move on to the next section and tackle our final project!

Final project – cement bag sorter

If you find yourself working on some type of bagging line, such as a line that fills cement bags, you will often need to know the lightest and heaviest bags that were produced. Typically, these statistics are used to give the overseers of the production lines a solid range of the amount of material in the bags. The overseers will often use this range as a quick check to ensure all the bags are in spec and nothing was over- or under-filled.

With this information about the project, we can now move on to laying out the requirements for it.

Requirements

The following are the requirements for the project:

1. For this project, we're going to assume that the customer is in a hurry and needs a quick patch that can implement the sorting behavior.

2. We can also assume that each production run is going to be small – that is, no more than four bags of cement per run.

3. The lightest and heaviest bags need to be highlighted.

Analysis

These requirements mean that we need an easy-to-implement algorithm. Also, since there will at most be four bags per run, the time complexity won't matter much so we can choose any easy-to-implement algorithm, regardless of time complexity. This means that a good candidate for this project is the bubble sort algorithm. For this project, we're going to sort all the bag weights in an array, sort the array, and assign the first element of the array (the lightest bag) to a variable called `lightest` and the last element in the array (the heaviest bag) to a variable called `heaviest`. The `heaviest` and `lightest` variables will be used to highlight those statistics.

Implementation

Based on what we concluded in the analysis, this project will be very straightforward. All we have to do is modify our bubble sort example. Now, to the inexperienced, reusing an existing example may seem like cheating; however, when it comes to algorithms, it is quite common to simply cut and paste the code from one project or example to another, as long as it doesn't violate more advanced rules that go beyond the scope of this book. However, with that being said, we can use the following logic:

```
PROGRAM PLC_PRG
VAR
     sizes : ARRAY [1..4] OF INT := [1,5,9,3];
     i, j, temp: INT;
     numOfBags : INT;
     lightest : INT;
     heaviest : INT;
END_VAR
```

In this case we have all the standard bubble sort variables (i, j) in an array that we're going to assume holds the weights of the cement bags. Next, we have a variable that will hold the overall number of bags made and a variable that will store the lightest bag made and the heaviest bag made.

The sorting logic will be as follows:

```
numOfBags := SIZEOF(sizes)/SIZEOF(sizes[1]);
FOR i := 1 TO numOfBags DO
     FOR j := 1 TO numOfBags - i DO
          IF sizes[j] > sizes[j + 1] THEN
               // Swap elements
               temp := sizes[j];
               sizes[j] := sizes[j + 1];
               sizes[j + 1] := temp;
          END_IF
     END_FOR
END_FOR

lightest := sizes[1];
heaviest := sizes[numOfBags];
```

The core of this logic is the bubble sort algorithm that will sort the bag weights. The first line of the program will calculate the number of bags in the production run and finally, the last two lines will retrieve the lightest and heaviest bags respectively. In short, after sorting, the first element will always be the smallest value and the last element - in this case - the total number of bags will be the heaviest.

When the program is executed, you should be met with *Figure 14.4*:

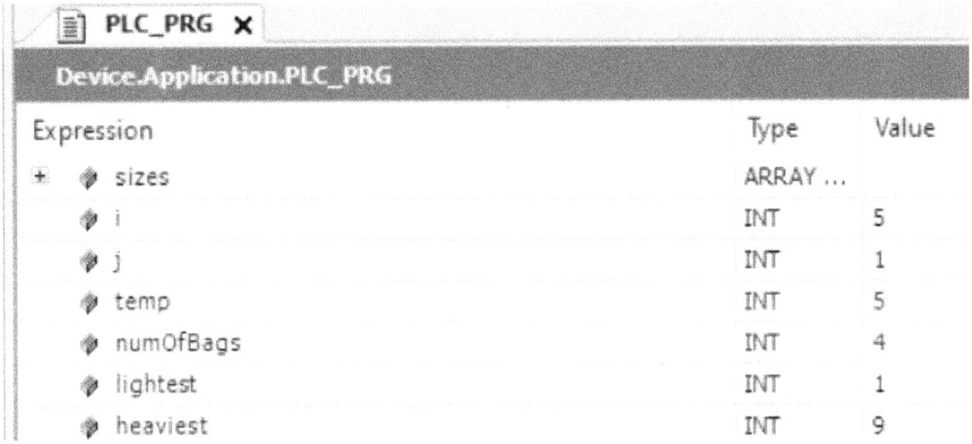

Expression	Type	Value
+ ⬦ sizes	ARRAY ...	
⬦ i	INT	5
⬦ j	INT	1
⬦ temp	INT	5
⬦ numOfBags	INT	4
⬦ lightest	INT	1
⬦ heaviest	INT	9

Figure 14.4 – The final project output

As you can see in the screenshot, the heaviest bag is 9 and the lightest bag is 1. These are the values that we would expect by simply analyzing the array. Now, this program is very generic in nature, as it should be for a real-world program. The program should be able to sort an array of any size, so whether we need to sort 3 bags of cement or 300, this program can accommodate that by simply adding more elements to the array.

For all intents and purposes, this bubble sort algorithm could be used in a real-world project if necessary. All one would have to do is modify the array accordingly. If one really wanted to get fancy, they could package the final project in a function and simply import the function into a real-world project!

Summary

In summary, this chapter has focused heavily on algorithms and how algorithms can be used for sorting. After reading this chapter, you should now be familiar with algorithms, metrics to denote an algorithm's efficiency, arrays, and more. As stated before, the key to this chapter is not memorizing the algorithms explored but understanding how to introduce yourself to algorithms and the necessary metrics and components to use them. In all, by now you should be able to Google an algorithm for a certain task and have the necessary experience to understand whether it will accomplish your task or not.

Now, as stated earlier in the chapter, there are algorithms for everything. Another area where algorithms are heavily used is in security. Security is a very important and often overlooked aspect of PLC programming. So, in the next chapter, we're going to explore how we can harden PLC against cyberattacks!

Questions

1. What is an algorithm?
2. How is pseudocode used with algorithms?
3. What does O(1) mean?
4. What is a more efficient time complexity: O(n*log(n)) or O(n^2)?
5. How does merge sort work?
6. Which is more efficient: bubble sort or merge sort?
7. For an array (0..238), how do you retrieve the first element in the array?
8. For an array (1..299), how do you retrieve the last element in the array?
9. How do you calculate the number of elements in an array?
10. Name three sorting algorithms.
11. Name three areas where an algorithm can be used.

Further reading

- *Bubble Sort – Data Structure and Algorithm Tutorials*:

  `https://www.geeksforgeeks.org/bubble-sort/`

- *Merge Sort – Data Structure and Algorithms Tutorials*:

  `https://www.geeksforgeeks.org/merge-sort/`

- *What is Big O Notation Explained: Space and Time Complexity*:

  `https://www.freecodecamp.org/news/big-o-notation-why-it-matters-and-why-it-doesnt-1674cfa8a23c/`

15
Secure PLC Programming – Stopping Cyberthreats

When it comes to **programmable logic controllers (PLC)** programming, cybersecurity is often not given a second thought. Many PLC programmers do not factor security into their software design. As such, automation software for PLCs, **human-machine interfaces (HMIs)**, and even full-blown **Supervisory Control and Data Acquisition (SCADA)** systems have recently become a major attack vector. One only has to look to recent history to see how sophisticated computer viruses have crippled PLC systems. For example, the infamous *Stuxnet* virus that destroyed Iranian PLC-controlled nuclear centrifuges or the Colonial Pipeline incident in 2021. Barring geopolitics and other reasons, these cyberattacks have proved beyond a shadow of a doubt that PLCs and automation software, in general, are now targets.

No matter what you're working on, cybersecurity is one of the most important aspects to consider when developing an automation system. When developing PLC software, one must remember that the software could be controlling heavy machinery, high voltage systems, or even doing critical tasks such as regulating power grids or manufacturing medicine. Imagine that you're working on a PLC project that controls crossing guards for railways. Consider that the software had a vulnerability in it and an attacker not only found it but exploited it. Suppose, for whatever twisted reason, they disabled the crossing guards so that they don't go down. In this scenario, drivers are blind to oncoming trains. Crossing the train tracks is now a potentially deadly venture. In short, people could easily die due to an unsecured PLC system.

Cybersecurity is not something that any programmer in any field should ever take lightly. Whole books, certifications, and college degrees are dedicated to securing systems, which, as you can deduce, means it's a complex topic. Nonetheless, this chapter is going to be dedicated to understanding and defending against basic attacks and exploring the basics of how you can harden your PLC-based systems. To do this, we're going to explore the following topics:

- What cybersecurity is and why it's important
- The basics of cybersecurity

- Common cyberattacks
- Attack prevention methods

To round out the chapter, we're going to create a simple lock-out program that can disable itself if a user inputs the wrong activation code too many times.

Now, this chapter is going to give an overview of the landscape for cybersecurity from a more theoretical point of view. Memorizing code examples will only get you so far in cybersecurity as each system will be written differently and, as such, will require different software patterns to defend the system. So, as you go through this chapter, ensure you are learning and understanding the concepts presented.

PLCs that run operating systems such as Windows or Linux distros can become prey to viruses. This is because, depending on how the malware is compiled or run, it may indeed be compatible with a sophisticated Windows or Linux PLC. An attack is much easier and more devastating to carry out on these systems. Keep this in mind as we progress through this chapter, most of the tips explored here are going to be geared toward protecting a network with PLCs on it as opposed to a single PLC.

Technical requirements

The code for this chapter can be found at the following URL:

`https://github.com/PacktPublishing/PLCs-for-Beginners`

As usual, you can download the code and modify it to gain a better understanding of how it works.

What cybersecurity is and why it's important

The goal of cybersecurity is to prevent unauthorized access to a network or system. In a more lay sense, cybersecurity aims to prevent hackers from breaking into a system and stealing data or using a system's functionality. In terms of automation, this could be someone who breaks into a system to tamper with it by shutting it down, damaging it by having it move uncontrollably, or any other nefarious action. Again, think of the railroad crossing example from the introduction. Put simply, cybersecurity is there to ensure that only the right people with the right intentions can access and control a machine or view its data.

Recently, cyberattacks have been on the rise. This increase has been due to many factors such as COVID-19, the rise of remote work, and, of course, **artificial intelligence** (**AI**). With the rise of digital systems, bad actors are increasingly using cyberattacks for financial gains, political/military (*Stuxnet*) advantages, or anything else that could give the cybercriminal some type of advantage. The alarming part is that the attacks are working. Military equipment and critical infrastructure have been damaged, as well as millions of dollars paid out to cybercriminals for the release of data and systems.

This is a very high-level explanation of what cybersecurity is. You've probably heard this spiel on the news a million times over. To really grasp why cybersecurity is important, we need to take a deep dive into the basics of digital security.

The basics of cybersecurity

Before we can start hardening our PLC systems from an attack, we first need to understand the basics. To start our exploration, we need to understand how bad actors can get into our system.

Vulnerabilities, threats, and risk

For a bad actor to get into a system, they need to find an entry point. Since they aren't supposed to have access to the system, they need to find a weak point that they can use to gain entry. This weak point is called a **vulnerability**. A day-to-day analogy of a vulnerability is like leaving your car unlocked. In terms of computer science, a vulnerability can best be thought of as a flaw in the software that leaves the software *unlocked* for someone to break into. For instance, a common flaw might be not locking out a user after they enter the wrong password three times or not encrypting a maintenance screen on an HMI panel. Flaws such as these can easily be exploited by bad actors to gain access and even exert control over a system. Just like leaving your car unlocked doesn't mean that a car thief is going to automatically steal your car, a security vulnerability doesn't mean that someone is going to break into your system. However, just like when there is a car thief in the neighborhood looking for unlocked cars, a security flaw can easily become a threat.

A **threat** is defined as a potential danger. For example, the car thief is a threat. A threat seeks to exploit a vulnerability; again, this is like a car thief looking for unlocked cars. So, essentially, a threat and a vulnerability are two halves of the same whole. If you remove the threat, you're safe, and if you remove the vulnerability, you're safe. As logic dictates, the trick to cybersecurity is to limit potential vulnerabilities so that threat actors cannot exploit them.

Suppose you find a vulnerability: a regular PLC programmer may panic; however, all hope is not lost. Once you become aware of a vulnerability, the first thing to do is either report or understand the **risk**. Essentially, the next step to take will depend on the organization you work for. For larger organizations that have more resources, there may be a dedicated team of experts such as a security team that is tasked with finding and eliminating issues or a development team that is responsible for the system's security. However, for many smaller organizations, this person might be you, and the first step in fixing the problem is understanding it. All vulnerabilities are not created equally. This means that some vulnerabilities may be very easy to exploit while others may be hard to exploit. Some vulnerabilities may cause a lot of damage if they are exploited, while others won't. Finally, there may be a high threat level for some vulnerabilities while others have a low threat level. Basically, the amount of risk involved with a vulnerability varies drastically.

All threats require a threat actor. In the next section, we're going to take a look at who could be attacking you and for what reasons!

Threat actors

For a vulnerability to become an issue, someone or something has to try to exploit it. In cybersecurity, this entity is called a threat actor. A threat actor is a person or group that tries to exploit a vulnerability in a system to gain unauthorized access to the system or data. In everyday lingo, these people are called hackers.

Hackers come in a variety of skill levels, ranging from nation-states (countries) that attempt to disrupt adversaries to unskilled computer enthusiasts who try to access systems for the thrill of it. No matter the intentions, as a software developer, your job is to write code in such a way that threat actors cannot gain a foothold in your system. Common types of threat actors are as follows:

- **Script kiddies**: Low-skilled hackers that utilize tools and techniques developed by others; in other words, beginner hackers.

- **Crackers**: This is what the term *hacker* typically refers to. A hacker, by definition, is just a computer enthusiast; however, the term got hijacked and is now synonymous with crackers, who are people who try to gain unauthorized access to a system.

- **Hacktivist**: A person or group that attempts to illegally access systems for a cause. An example of a hacktivist group is the *Anonymous* group. In terms of automation, threats could stem from groups that are against a given industry.

- **Nation-state**: Usually, a government or government agency with access to advanced resources. The attacker(s) can pull off sophisticated attacks that are hard or nearly impossible to defend against. In terms of automation- and PLC-based systems, this is a prevalent threat. For example, many power grids, oil pipelines, and other critical infrastructure are prime targets for cyber terrorism and cyber warfare in general.

- **Insider threat**: A person or group that has been given access to a system that is now using that authorization for nefarious purposes, intentionally or not. This is a very common threat in the automation world. It is common for insider threats to be unintentional. For example, an employee could click on a malicious link, install a program that contains malware, or accidentally do any number of things that could adversely impact or cause damage to a system.

- **Competitor**: This is typically a rival organization that illegally accesses a target organization, usually for economic gain, by accessing sensitive data or disrupting processes. This is another major threat to automation systems.

The list goes on. There are many more types of threat actors out there. However, these are the most common. The first step to protect a system from a security threat is to consider **authentication, authorization, and accounting (AAA)**!

Exploring AAA

The first thing that a developer can do to minimize threats and vulnerabilities is to employ AAA. The following is a high-level overview of AAA.

Understanding authentication

As the name suggests, **authentication** means the user or operator has to prove they are who they say they are. In its simplest form, this is done with a username and password; however, modern security advances now allow for biometrics, facial recognition, and **multi-factor authentication** (**MFA**) to help enhance security. When the operator inputs the correct username and password combination along with any other additional security features, they are granted access to the machine or data. For automation, this is especially useful. For example, you don't want an untrained operator toying around with calibration data, and you don't want a random person to press a button and start a machine. Typically, it is a good idea to assign each machine operator a username and password that they can use to access the machine's operational controls. Due to the air-gapped (not connected to the internet) nature of machines, this can sometimes be less than optimal to do. When a user cannot be easily added to a machine or group of machines, many will opt for the less-than-secure method of shared accounts. Sometimes, companies will create an account such as `operator` or `admin` and give the necessary credentials to only trusted users. These shared accounts are not recommended and are very insecure and dangerous, but it is very common to see these accounts in the automation world. Regardless of the type of account, once a user has been authenticated, the system needs to determine what they're allowed to do.

Understanding authorization

Where authentication is accessing a system, **authorization** is permitting a user to perform an action. In terms of automation, it is common to only allow certain users to access certain features of a machine. For example, only maintenance technicians should be allowed to enter calibration data, while only a production engineer should be allowed to alter production parameters. To accomplish this, certain accounts will come with certain privileges. These privileges typically come packaged with a user's account.

Usually, users are placed into what are called **groups**. A group is a collection of users that have the same privileges in a system. Essentially, when a user logs in to the system, they will authenticate their credentials. After verifying the user, the system will then determine which group a user is placed in and determine what they can do and what data they can access. Creating and adding users to groups is basically a shortcut to having to assign the same permissions to multiple individuals. Now, creating groups can be tricky on a traditional PLC. To pull this off, you'll need to implement either an advanced SCADA system, use a general-purpose programming language to create an HMI that lives on an actual computer, or use an advanced PLC that can utilize complex logic such as a high-end Beckhoff, Siemens, Rockwall, or other advanced PLC brand.

For most PLC projects that are not connected to a network, creating a group or assigning individual permission on multiple machines is going to be impractical. This is why many industrial settings use shared accounts. In the case of shared accounts, a manager or other person will simply give certain people a shared username and password that have the appropriate levels of permissions. For many applications, this will typically work; however, it makes the final "A" very difficult to implement.

Understanding accounting

Accounting is easy to understand. In short, accounting is tracking who logged in to a system and what they did. What is logged will vary greatly; however, at the very minimum, you should log the following:

- **Who logged in to the system?**: This will usually include things such as users, IP addresses, and location details.

- **How long was the person logged in to the system?**: Typically, you want to log when the person signed in to the system and when they signed out.

- **What did the person do?**: For critical systems, it is vital to log who did what and when. For example, if person X logged in to the system and adjusted the flow of oil through a pipeline and the pipeline breaks, there is a clear record of who did it. Once the person is identified, they can be questioned as to the circumstance in terms of why the action was taken.

When it comes to machines such as everyday manufacturing equipment that is not connected to a network and shared user accounts are used, accounting may not be totally necessary. When it comes to cases such as these, it will be almost impossible to accurately tell who logged in and performed whatever action, rendering accounting inaccurate and unnecessary. However, if you're working on critical infrastructure such as power grids, important oil pipelines, railroads, pharmaceutical systems, or anything of the sort, accounting and unique user accounts are mandatory.

When it comes to critical applications, accounting is of vital importance. Having accurate records of who logged in to the system, where they logged in from, and, most importantly, what they did while logged in is critical to the safety and security of the applications the software governs. This information is important for cyber forensics because if someone does break into a critical system and causes severe damage or death, the logs will at the very least have some information to help catch the perpetrator. Now, as stated before, if this is a simple machine that is not in control of critical infrastructure, accounting may not be necessary. However, some industries do require by law that proper accounting is implemented, so ensure you know what laws and guidelines govern the project. With AAA understood, let's move on to another vital concept: air-gapped systems!

Air-gapped systems

Air-gapped systems are simply systems that are not connected to the internet. Since air-gapped systems are not connected to the internet, they are hard, if not impossible, to hack. An air-gapped system may be connected to other systems; however, the overall system is disconnected from the raw internet. In more standard automation, such as machines for traditional manufacturing environments, many systems are going to be air-gapped. This isn't necessarily for security reasons, as air-gapped systems are often easier to make.

Just because a system is air-gapped, it doesn't mean that it can't be infected. A general rule of thumb is no system is hackproof. In the following section, we're going to explore cyberattacks and how bad actors can infiltrate systems.

Common cyberattacks

Cyberattacks come in many shapes and sizes. There are many, many ways to break into a digital system and either take full control of the machine, steal data, or perform other malicious actions. The following is going to be a rundown of how some basic cyberattacks are pulled off and how to defend against them. The first area that we're going to explore is **information gathering**.

Exploring information gathering

The first step in pulling off an attack is understanding your target. In today's day and age, this has never been easier. With the widespread adoption of social media, gathering information on a target is very easy and, more importantly, legal. Information gathering is used to scout people who may have access to a specific system in an attempt to gain information that can be used nefariously to access the system. Suppose a hacker wants access to an oil pipeline that's controlled by *Really Sweet Oil Co*. A common tactic is to scout out LinkedIn for active employees of the company. Once the attacker identifies a person who may have access to the target system, they'll attempt to gather information about them. Usually, they gather as much information about their target as possible, such as the technologies they work with, contact information, common themes, and so on. They'll also explore other social media sites such as Facebook, Instagram, and so on to gather personal information such as the following:

- Special dates such as birthdays or anniversaries
- Names such as pet's names, spouses/significant other's names, maiden names, children's names, and the like
- Addresses
- Contact information
- Blackmail information

The purpose behind gathering this information is to build a profile of the target. Once the profile is built, the attacker can then use that information to try to find vulnerabilities so that they can infiltrate the system. Another concept that can loosely be thought of as information gathering is social engineering.

Exploring social engineering

The whole concept of social engineering is tricking a victim into revealing sensitive data or doing something that is against their best interest, such as clicking a malicious link. Social engineering comes in many forms, with the most common being phishing. However, other social engineering tactics can include the following:

- Spear phishing
- Scareware
- Pharming

There are many more tactics. There is a lot to social engineering, and a single book section will not do it justice. So, to get our feet wet, this section is going to focus on one of the most common tactics: phishing!

Understanding phishing

If you've ever been contacted by a royal family member who's asking you to click a link to redeem King Tut's hidden treasure, you've encountered a phishing attempt. Out of all the social engineering methods that attackers employ, phishing is arguably the most common and, in many cases, the most effective. To summarize what phishing is, it is simply some type of malicious email, text, social media communication, or other digital communication that attempts to trick you into entering sensitive information that will be sent back to the attacker. The message will usually come disguised as a link to a popular bank brand, a credit card organization, the **Internal Revenue Service** (**IRS**), or anything that looks legitimate. If clicked, the link will usually route you to a login form or something of the sort. It will usually prompt you to enter your login info, which will be sent back to the attacker. An example of a mock phishing email can be viewed in *Figure 15.1*:

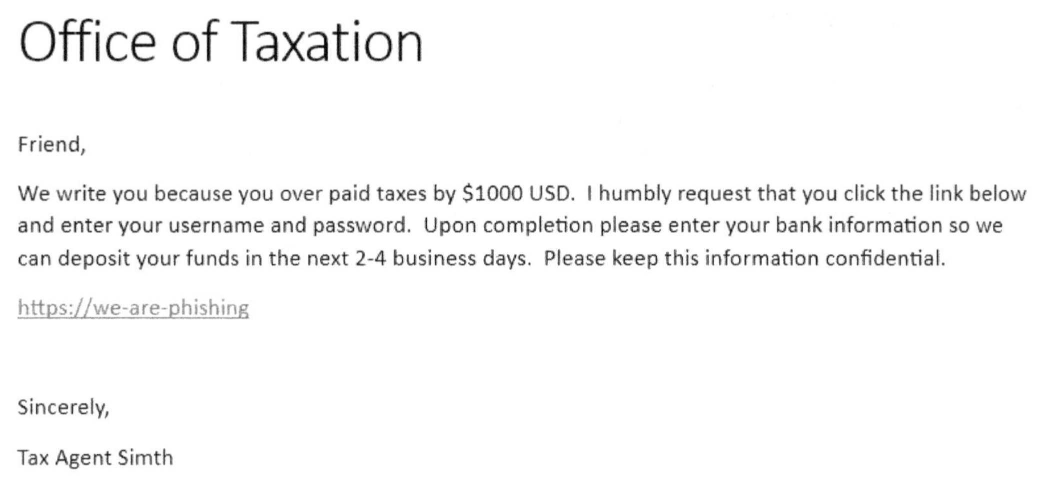

Office of Taxation

Friend,

We write you because you over paid taxes by $1000 USD. I humbly request that you click the link below and enter your username and password. Upon completion please enter your bank information so we can deposit your funds in the next 2-4 business days. Please keep this information confidential.

https://we-are-phishing

Sincerely,

Tax Agent Simth

Figure 15.1 – Mock phishing email

This is a mock email whose goal is to make you click the link. Notice the spelling and awkward wording. Though phishing attacks are becoming more sophisticated, bad and awkward wording is still common in these attacks. Contrary to popular belief, bad spelling and grammar do not stem from a poor understanding of the English language. Attackers will usually do this to test the victim. If the victim still goes along with the email, the attacker/scammer knows that the person is going to be more easily duped into providing the information they are after. Also, notice the link; it is clearly not an IRS or tax office link. Illegitimate links will often be nonsensical links or strings of random characters. In general, if you get a message that includes things such as promises of large sums of money, money that you didn't know you were owed/owe, love requests, bad grammar, nonsensical email addresses/links, or so on, it's, unfortunately, a phishing attempt!

What does this threat mean in terms of automation? Social engineers will often send messages to employee email addresses or work phones. These messages will look legit, and they will attempt to get users to input information such as their login credentials and so on. If the system can be accessed remotely, the attacker will usually try to do so with the stolen credentials. This means that any system the stolen credentials have access to can be accessed by the perpetrator. If you're working with air-gapped machines such as individual welders, bagger systems, and so on, you won't have much of a problem. However, if your system is not air-gapped, the attacker can cause all kinds of havoc on your system or network.

This is just one form of social engineering, and, as with many other concepts explored so far, whole books have been dedicated to the art of social engineering. For now, it is enough to be familiar with the concept and phishing. We will now switch gears and explore some common ways to crack a password.

Exploring password hacking

When one thinks of hacking, one will often conjure up the notion of stealing passwords. Unlike most stereotypes, this one is very often true. Many common attacks are designed to guess or crack a password. PLC-based systems are often more vulnerable to this kind of attack because it is common to have system functionalities protected with a very weak password and no username. Therefore, the best way to defend against a password-cracking attack is to understand how to implement a password-cracking attack. The first attack we're going to look at is the password-guessing technique!

Understanding password guessing

As the name suggests, **password guessing** is the act of trying to guess a password. Simply put, password guessing is where an attacker inputs different passwords until one works. Password guessing isn't just inputting random passwords until one works. Typically, an attacker will gather information about the target before attempting this. The attacker will attempt to use common passwords, a combination of information about the target, and more to try to generate a correct password. In all, password guessing is typically a very low-tech solution to password cracking. This attack can be automated with specialized scripts and programs or carried out manually. Due to the simplicity of the attack, it can be carried out by anyone, but it is commonly performed by script kiddies. A more advanced derivative of this attack is called a dictionary attack.

Understanding dictionary attacks

A **dictionary attack** is a more sophisticated and automated version of a password-guessing attack. A dictionary attack uses a list of passwords and password combinations along with a password-cracker program such as John the Ripper. For these attacks, the password-cracker program will read from the dictionary and run those passwords against the system. These dictionaries are text files that are gigabytes in size, so they contain massive amounts of passwords. If you think of how fast a computer is, it can test hundreds, if not thousands, of passwords a second. These attacks are much more effective and faster than password-guessing attacks as more passwords can be tried out in a much shorter amount of time. There is another derivative of this attack that does not use a dictionary called a brute force attack.

Understanding brute force attacks

A **brute force attack**, on the surface, is very similar to a dictionary attack. They are both automated attacks that can try many passwords in a short amount of time. However, where a dictionary attack uses a password list, a brute force attack does not. A brute force attack will try every different combination of characters until it finds one that works. Brute force attacks are not limited to the passwords in a file. In theory, a brute force attack can be carried out indefinitely until it finds a password that works. Now that we understand password cracking, we need to move and explore the different types of malware programs.

Malware

No conversation on cyberthreats is complete without exploring **malware**. In its most simplistic sense, malware is malicious software. Malicious programs come in many different shapes, forms, and functionalities. The following are the most common forms of malware, along with what they do:

- **Trojan Horse**: A Trojan Horse, or simply a Trojan, is a program that masquerades as a legitimate program that tries to trick the victim into installing it. Once installed, the program will implement its malicious payload and do whatever it was designed to do.

- **Worms**: Worms are self-replicating programs. Once installed, the malicious program will start to reproduce itself and spread to other connected systems.

- **Rootkit**: A program that gives admin-level access to the attacker. Admin-level access is basically the highest level of permissions a user can obtain on a system. As such, once an attacker has this level of access, they can essentially do whatever they want with the system. These are exceptionally dangerous malware infections.

- **Virus**: A catch-all term that is used by most to refer to malware. By definition, a virus is a program that behaves like a biological virus. These programs infect other files and programs and cause all kinds of issues on a host system.

- **Keyloggers**: A keylogger is a program that, when installed on a victim machine, will record keystrokes and send them back to the attacker. These programs are commonly used to capture sensitive information such as usernames and passwords.

There are many more types of malware programs, such as backdoors, spyware, adware, and so on. In practice, malware programs will often exhibit behaviors of many of these program types. In all, malware is usually not a one-size-fits-all thing. Malware programs are often complex and sophisticated pieces of software that have many attributes from many different categories of the aforementioned list.

There are many more types of cyberattacks that can be pulled off and malware programs that can cripple a system. The attacks and malware listed are just some common ones. However, since these are very common, they are easy to defend against.

Attack prevention methods

Preventing cyberattacks starts with understanding access controls. Traditionally, there are three categories of access controls:

- **Technical**: Software and hardware that help prevent cyberattacks. This could be software as simple as implementing password protection for sensitive controls and data to more advanced techniques such as implementing firewalls.

- **Administrative**: Things such as policies and procedures that are not technology-based but help prevent cyberattacks. Administrative controls can be company policies such as preventing shared user accounts or requiring mandatory cybersecurity training for operators.

- **Physical**: As the name suggests, physical controls are security measures such as fencing, security guards, access cards, and so on.

Depending on what you're working on, you may only need technical or a combination of administrative and technical controls. Physical controls are only necessary when the project is critical in nature or high value, such as an oil pipeline, military factory, smart factor, or something along those lines. For an average factory environment, technical and administrative controls will typically suffice. This section is mostly going to focus on technical controls. However, the first cyber defense we're going to look at is how to defend against social engineering.

Stopping social engineering

Okay – so, you can't stop social engineering; unfortunately, the attacker has a vote in whether they're going to attack you or not. However, there are administrative controls that can help protect not only you but your organization as well. When you think about what social engineering is and how it is carried out, you will find you can defend against it with both administrative and technical controls.

Since a common vector for social engineering is digital communication such as emails or SMS, there are a couple of controls we can put in place to thwart would-be attackers. For starters, to stop phishing attacks, companies can invest heavily in spam blockers. In terms of administrative controls, an organization can require all employees who have a company cell phone to block calls and messages from unknown callers. Organizations can also invest heavily in cybersecurity training and auditing. In short, organizations should require all employees to take cybersecurity training on a regular basis to educate them on how to spot social engineering attacks. At the same time, companies should routinely send phishing emails to employees to ensure they understand the training. If the employee fails, they should be required to take an intermittent refresher course. Is this going to be a foolproof way of preventing social engineering attacks? Well, no! However, the first step in preventing an issue is understanding the issue. So, what about password-cracking attacks? How can one defend against those?

Defending against password crackers

The first rule in mounting a cyber defense is not to give an attacker a clue. In the early days of the internet, it was very common to put a prompt under a password box that said something along the lines of, **Please Enter Your Six Character Password**. This gave attackers an obvious clue about the password length. They knew that at the very least, the password was going to be six characters long. As such, when you're creating an HMI or writing a PLC program, never give a clue to the possible number of characters in a password. Outside of that, we can set password sizes behind the scenes.

Password length

In technical lingo, the length of a password refers to the number of characters in a password. The longer the password is, the harder the password will be to crack. In short, a long password is going to be much harder to guess, and it will take a lot longer for a brute force attack to figure out. Today, most passwords should be at least eight characters long. This can be done very easily in CODESYS. In short, we can use the following code:

```
PROGRAM PLC_PRG
VAR
     password      : STRING(255) := 'password';
     length        : UINT;
     acceptPass    : bool;
END_VAR
```

In this variable block, the `password` variable will hold the password, the `length` variable will hold the length of the password, and the `acceptPass` variable will toggle to `True` if the password is at least eight characters long and `False` if it is not. The logic is as follows:

```
length := LEN(password);
IF length >= 8 THEN
     acceptPass := TRUE;
ELSE
     acceptPass := FALSE;
END_IF
```

The key to this program is the LEN function. This function will return the length of the password. If the password is at least eight characters long, the program will set `acceptPass` to `True`; else, it will set it to `False`.

When the program is executed with the default password value set to `password`, you should be met with the output seen in *Figure 15.2*:

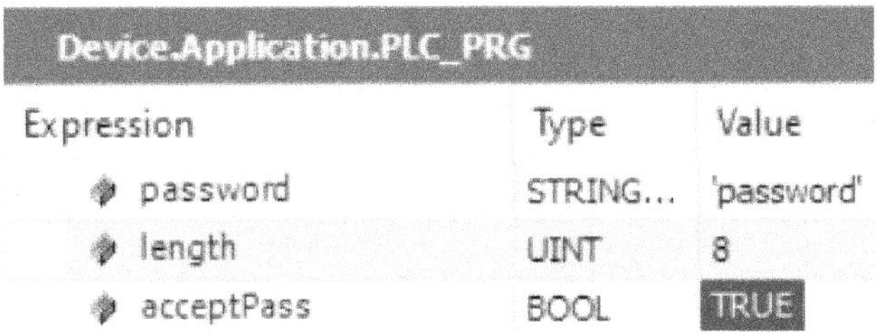

Device.Application.PLC_PRG		
Expression	Type	Value
password	STRING...	'password'
length	UINT	8
acceptPass	BOOL	TRUE

Figure 15.2 – Accepted password

If you were to set the `password` variable to something such as `passw`, you should be met with the output seen in *Figure 15.3*:

Figure 15.3 – Rejected password

Even if a password is eight characters or more, this still doesn't mean that the password is strong or safe. Most cybersecurity experts will now say that instead of using a password, a user should use a passphrase. That is, instead of using Fluffy for a password, a user should use FluffyIsMyDog. A hacker would have a much tougher time cracking the lateral. Most experts also agree that all passwords or phrases should include numbers and special characters. For example, an even better password would be FluffyIs@MyD0G13*. This passphrase would be even tougher to crack.

In terms of development, a general rule of thumb is to keep your HMI as *dumb* as possible. This essentially means you want your HMI to do as little computing as possible. However, checking strings for complexity can be difficult in a PLC program. This is mostly because few utilities can be used to easily check complexity. Also, if you consider how PLCs work, it is very easy to download and upload the PLC program if the PLC does not prevent the code from being downloaded without a key. This means that for certain PLCs, if someone were to be able to download the code, they could get password information, get clues on what passwords include, or bypass security controls altogether. When you consider these facts, it makes much more sense to put the security logic in the HMI, if possible, especially if the PLC code cannot be locked out. If you are using a general-purpose programming language such as C#, Java, C++, or something similar, it will be much harder for an attacker to get their hands on the HMI binaries and reverse-compile them, and those programming languages often have built-in security utilities. Even if you're using a canned HMI programming system or SCADA system, there will usually be features that will allow developers to set password characteristics and protect HMI screens that will lock out the machine's functionality.

> **Note**
> There are a lot of opinions about where one should store a password if the system is simple and does not contain storage components such as databases. Storing the password in the PLC can offer more control and ease of access for troubleshooting; however, storing the password in the HMI is technically safer.

Storing a password in an HMI or a PLC program is never a good idea. At best, this should be considered a necessary evil. When it comes to storing passwords, the best solution is to encrypt the password and store it somewhere safe, such as in a database. However, for many machines, especially air-gapped systems, this may not always be possible.

> **Note**
> The safest way to store a password is typically in some type of storage system such as a database. It is often required that the password is hashed before it is stored. Hashing a password is a mathematical operation on a string of text that will transform it into a fixed-length string of alphanumeric characters, which will obfuscate the true meaning of the password.

The next prevention method that we need to explore is pen testing.

Pen testing

Another common technique that can help protect systems is pen testing. Pen testing is an advanced cybersecurity auditing technique that is used to find vulnerabilities in a system. This auditing technique is, essentially, hacking a system; however, instead of hacking for some type of nefarious gain, pen testing attempts to find vulnerabilities in a system so that they can be fixed.

Pen testing can be very complicated and will usually need dedicated resources such as people well versed in pulling off various types of attacks. For smaller organizations that do not have the resources to have a pen tester, it will often be up to the dev team to do the testing. For such organizations, pen testing will often be overlooked; however, if possible, it is wise to become familiar with conducting basic cyberattacks and try to use them on a system that is under development.

A quality pen-test audit will usually require specific technologies that are specifically designed for breaking into systems, such as the Kali operating system, which may not be available in all organizations. However, very basic penetration techniques can be performed manually. For example, cracking a password to test its strength can be done manually using the password-cracking technique mentioned previously.

To get some practice with this rudimentary pen-testing technique, let's try to manually crack a password. To practice this, download the `Pen Test` program from the GitHub link in the `Chapter 15` folder. Without examining the code, try to guess the password. If you guess the password correctly, the `accessGranted` variable should toggle to `True`. As a hint, the password is a common dog's name that starts with `Foo`, ends with `fy`, and the developer's favorite number, which is `123`.

In the real world, it is best to give no hints to the person trying to crack the password. This pen-testing example is just to give a little hands-on practice. In the real world, it is best to use pen-testing tools; however, using those tools, especially for testing automation software, is beyond the scope of this book. Nonetheless, if you're working for a small organization and the passwords do not protect anything super critical and are only protecting things such as HMI screens or the like, trying to manually crack the passwords will have to do.

Before we move on to the final project and explore the ultimate way to defend against password-cracking attacks, we need to explore how to defend against malware.

Malware defense

Malware can be a problem on any system, even an air-gapped machine. However, there are some things you can do to protect yourself and your machines. The first and most obvious is to be cautious about what you plug into your USB ports!

Exploring USB control

USB drives can be vectors of cyber disease. If you're working with a PLC that supports USB devices or a PLC that is connected to a system that supports USB devices, you will need to implement strong administrative, physical, and technical controls to limit their use. First and foremost, an organization should put policies in place that prevent any unapproved USB from being plugged into a PLC or network. USBs of unknown origins, such as USBs that were found, should never be allowed to be plugged into a PLC or network. In short, USB drives from unknown origins can easily house malware that can be installed on the system as soon as it is plugged in.

Since USB drives are vital for data transfer and maintenance, what should one do? The first thing an organization should do is heavily monitor the USB ports of a network or system. This could include locking out unused USB ports, using monitoring software and the like. This should be coupled with an administrative policy that dictates that only authorized USBs can be plugged into the system. There should only be a few authorized USB drives that are guarded by a trusted person such as a manager. The drives should be in a secure location that is constantly locked. When someone needs to use one of the drives, there should be a sign-in sheet that includes information such as the following:

- Who is checking the drive out?

- At what time are they checking the drive out?

- Why are they checking the drive out?

- Work area where they are using the drive

- When did they check the driver back in?

Checkout sheets such as these can be vital in figuring out when and where an infection originated from if malware is introduced to the system. With this information, the malware can be easily pinpointed, and counteractions can be quickly taken to prevent the spread to other machines. Another vital defense that can be taken by organizations to prevent the spread of malware is air gapping when possible!

Air gapping when possible

Whenever possible, machines and networks should be air-gapped. Air gapping has pros and cons, but in terms of security, it is usually a good idea to air gap a system. As stated before, air-gapped machines cannot reach the raw internet. This means it is very difficult to attack and introduce malware from outside the network. If malware is ever introduced into the network or system, it is already quarantined to only connected devices. Another effective way to prevent malware is an old-fashioned anti-virus program.

Understanding anti-virus software

Anti-virus software, as you can guess, will remove malicious software such as viruses from a computer. Of course, this is not an optimal solution for many PLCs; however, these can be used to great effect on PLCs that run a common operating system or on PC-controlled systems. If the controller does not use a typical operating system, it will be impossible to run a normal anti-virus on the device; however, for a PLC such as this, a virus is going to have to be specifically designed to attack that brand and, in many cases, model of PLC. Therefore, unless it is a virus such as *Stuxnet*, there is a low probability of anything happening. Typically, malware will not attack a PLC. Instead, a malware program will attack the network that a PLC is attached to and send erroneous commands to the PLC. So, if your PLC is attached to a control network that is Linux- or Windows-based, you want to ensure that you have adequate anti-virus software protecting the devices on the network.

There are many other ways to protect against malware. These few topics are just some high-level, common-sense defenses to get you thinking. Whole books have been dedicated to malware defense, and it is advisable for any PLC developer to explore this topic more. With that, we're going to move on to our final project and create a simple PLC-based login system!

Final project – a PLC-based activation system

As stated before, you typically don't want your PLC to be responsible for handling login and account maintenance, mostly due to how easy it is to download PLC code and bypass any security systems. A common example of this is customers trying to bypass activation codes. PLC-based equipment is often very costly, and organizations that build machines will often allow the customer to make payments on the machine. However, some customers are less than ethical in this area. It is not that uncommon for a customer to try to guess the activation password to avoid paying for the rest of the machine. This means as developers, we have to be clever and put in safeguards to prevent this.

Design

When you think about it, cracking an activation code is the same as cracking a password. Simply put, a person can use the exact same techniques. They could guess the password, use a brute force attack, or if they have the necessary experience, they could try to use a dictionary attack with a custom dictionary. No matter which attacks they use, they all have the same main weakness: it will usually take multiple attempts to guess the correct password. Unless the attacker is very lucky, each attack will require multiple attempts to crack the activation code. We can use this to our advantage when we create the activation software. All we must do to defeat any password-cracking attempt is to limit the number of password attempts!

For this program, we're going to give the user three chances to input the correct activation code. After three attempts, we're going to assume that the user is trying to guess the code and we're going to lock the machine out. To do this, we're going to use the loop nature of the PLC and implement a program based on the following pseudocode:

```
Code = 0
Password = 8869
Count = 1;
Activate = false
Lockout = false

Input code

If count > 3 then
    Lockout = true
End_if

If code <> 0 then
    If code <> 8869 then
        Count = Count + 1;
        Code = 0;
    Else
        Activate = true
    End_if
End_if
```

To implement this code, start with the following variables:

```
PROGRAM PLC_PRG
VAR
    code            : INT  := 0;
    count           : INT  := 1;
    activate      : BOOL := FALSE;
    lockout       : BOOL := FALSE;
END_VAR
```

For this program, the code variable will hold the access code that the user inputs, the count variable will keep track of the number of activation attempts made, the activation variable will determine if the machine is activated, and finally, the lockout variable will determine if the machine is locked out or not.

The main logic for this program will mirror the pseudocode and be as follows:

```
IF count > 3 THEN
    lockout := TRUE;
END_IF

IF code <> 0 THEN
    IF code <> 8869 THEN
        count := count + 1;
        code := 0;
    ELSE
        activate := TRUE;
    END_IF
END_IF
```

For this code, the first thing the program will do is check the number of attempts made. We're going to assume that there is at least one check to account for the user's first activation attempt. Each time the program iterates, it will perform this check first. If this check fails (the user still has attempts left), it will move on to check the activation code. If there are no attempts left, it will lock the machine out by setting the lockout variable to True.

The second check will be to ensure the code is not 0. This logic prevents a runaway code situation by ensuring the user inputs data into that field. The nested IF statement will check if the codes are a mismatch. If they are, it will increment our attempts counter and reset our code to 0. If the codes match, the machine will activate by setting the activate variable to True. To test this, write 1111 to the code variable three times. You should be met with the output seen in *Figure 15.4*:

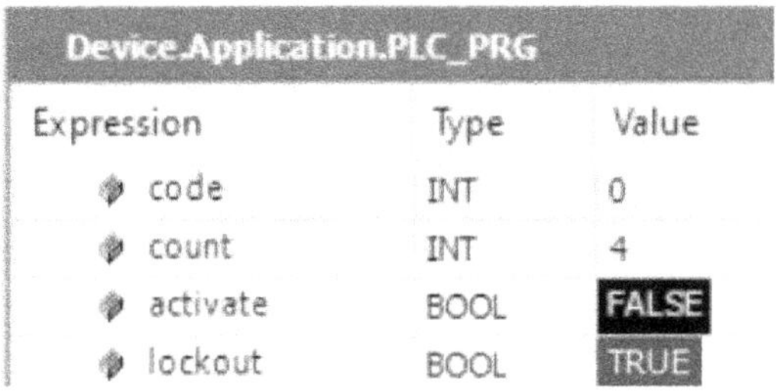

Device.Application.PLC_PRG		
Expression	Type	Value
⬦ code	INT	0
⬦ count	INT	4
⬦ activate	BOOL	FALSE
⬦ lockout	BOOL	TRUE

Figure 15.4 – Machine locked out

Once you get to this point, reset count to 1 and lockout to False. This time, enter 8869 for the code, and you should be met with the output seen in *Figure 15.5*:

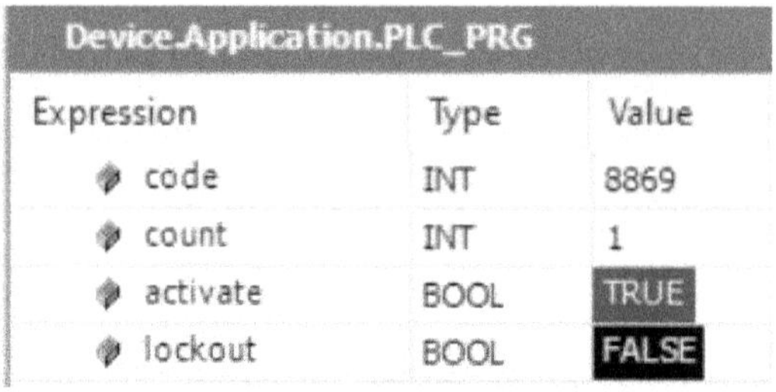

Figure 15.5 – Activated machine

In this case, since we input the correct code, the machine activated itself. Now, as stated before, having this code in the PLC isn't ideal. However, depending on the HMI system that you're using, network setup, and so on, you may have to code the logic in the PLC. If you can avoid doing this, you should; if you can't, then it's better than nothing. Regardless, the code presented can be easily ported over to any programming system you use. As such, you can always reference this if you need to lock out a real-world machine.

Certifications

In the IT world, security certifications are important. As an automation professional, it would be advisable to seek out a few certifications to help you grow and bulk up your cyber defenses. The following is a short list of some good certifications that you should investigate if you feel you would like to pursue cybersecurity more:

- **Linux Professional Institute (LPI)** Security Essentials
- CompTIA Security+
- **CompTIA Security Analyst (CySA+)**
- CompTIA PenTest+
- **International Information System Security Certification Consortium Certified Information Systems Security Professional (ISC2 CISSP)**

These are just a few common IT security certs. You don't technically need them, but they are good learning resources for mounting cyber defenses. Though these are geared toward traditional IT systems, they cover material that can be used to take the security of your PLC systems to the next level!

Summary

In summary, this has been a crash course in cybersecurity for automation systems. We've explored many security-related topics such as common attacks, defenses, malware, and more. Security is a very complex topic, and many aspects go well beyond the scope of this book. Nonetheless, this chapter has laid down the basics of cybersecurity; in short, security starts with this chapter. With security established, we need to move on to troubleshooting systems!

Questions

1. Is the activation code in the final project a weak password?

2. What would be a better activation code for the final project?

3. What is social engineering?

4. What is a brute force attack?

5. What is a dictionary attack?

6. What is an air-gapped system?

7. Should you air-gap a system?

8. What are the triple As of security?

9. What is a threat?

10. What is a vulnerability?

11. If you find a USB drive, should you plug it into your network? Why/why not?

12. Can a PLC be infected with malware?

13. Should you secure a PLC or the network the PLC is on?

14. Should you store passwords in the PLC or HMI, considering neither is attached to a network?

15. What is an insider threat?

16. What is a script kiddy?

17. Who can be a hacker?

18. What is the difference between a hacker and a cracker?

19. What is a hacktivist?

20. What is a nation-state attacker?

Further reading

- *Threat Actors*:

  https://www.sophos.com/en-us/cybersecurity-explained/threat-actors

- *5 Password Cracking Techniques Used in Cyber Attacks*:

  https://www.proofpoint.com/us/blog/information-protection/password-cracking-techniques-used-in-cyber-attacks#:~:text=Password%20cracking%20typically%20refers%20to,access%20to%20systems%20and%20resources

- *The 12 Most Common Types of Malware*:

  https://www.crowdstrike.com/cybersecurity-101/malware/types-of-malware/

- *What is social engineering?*:

  https://www.ibm.com/topics/social-engineering#:~:text=Social%20engineering%20attacks%20manipulate%20people,their%20personal%20or%20organizational%20security

16

Troubleshooting PLCs – Fixing Issues

One of the most common tasks a PLC programmer is going to perform is troubleshooting issues. This might be troubleshooting issues related to software or related to hardware. No matter how old a machine is, how complex it is, or how sophisticated it is, there are going to be problems that prevent it from carrying out its duties.

PLC-based systems, or, for that matter, any system that has software and hardware components, can be very tricky to troubleshoot, especially when the system is new or there have been modifications to it. There are an infinite number of issues that can cripple or, at the very least, hinder a machine. Unfortunately, there is no one-size-fits-all solution; however, there are common issues that can arise that will present similar behaviors.

This chapter will explore common issues that can arise in a PLC-based system. The chapter will also provide an overview of how to diagnose them and repair them. To do so, this chapter will cover the following:

- Common causes of PLC issues

- Hardware-related issues

- Troubleshooting techniques

Finally, to round out the chapter, we're going to troubleshoot a theoretical issue causing intermittent issues that are causing the PLC to deviate from its normal behavior.

Now, there is no silver bullet to troubleshooting a piece of hardware or software. Though common tips and tricks will be presented, the goal of this chapter is to present you with a mindset of troubleshooting issues. As such, think of this chapter as more of a framework as opposed to prebuilt patterns.

Technical requirements

This chapter is going to be focused on troubleshooting. This means we're not going to develop any code in this chapter. Instead, this chapter is going to use prebuilt code that we're going to troubleshoot. Before you proceed, please download the code for this chapter:

`https://github.com/PacktPublishing/PLCs-for-Beginners`

Also, a solid understanding of *Chapter 2*, is required before you can tackle this chapter.

Common causes of PLC issues

PLC issues come in all shapes and sizes. However, there are factors such as age, environment, and so on that will contribute to the demise of a system. To begin the discussion, we're going to talk about the all-too-common broken software problem.

Broken software

If you've worked as a PLC programmer for more than a few weeks, you've probably heard someone complaining about broken software, especially if you work directly with third-party customers. Now, it is nearly impossible to break software. Broken software is usually the result of malfunctioning hardware, or more commonly, user error. Typically, what a layperson will perceive as broken software will stem from operator errors. It is not uncommon for new or poorly trained operators to use a machine in an unattended way. This will almost certainly result in a bizarre behavior.

The other most common cause for broken software is hardware malfunctions. Typically, software is designed to read inputs from hardware such as sensors and send commands to other types of hardware such as valves, motors, and so on. If a hardware component starts to fail, it will usually behave in unusual ways. For example, valves may not open or close in their expected manner, motor drives will not respond as they should, and so on. It is very common for small problems in the hardware to mimic malfunctioning software. Sometimes, these problems are so small that if a technician is not experienced with malfunctioning hardware or the nature of software, they will often mistake the issue for faulty software.

As we will explore later, there are some causes that could result in software malfunctions. However, these issues are usually not due to issues with source code; instead, these issues are caused by the PLC itself. Before we get into determining whether there is a programmatic issue, we need to look at how the environment can disrupt the PLC's operations.

Exploring environmental issues

It's no secret that most PLCs will operate in very harsh conditions. These environments can have extreme temperatures, extreme temperature fluctuations, adverse weather conditions, or be exposed to extreme humidity or moisture. It's important to remember that even the most basic PLC is a computer, and moreover, an electrical device. One of the silent killers of PLC is temperature.

How temperature affects your PLC

Believe it or not, one of the most crippling environmental factors that a PLC can face is temperature. All electronic devices have an optimal temperature range. If the device is exposed to temperatures outside of its normal operating range for a long time, it will begin to behave erratically and then eventually fail. Normally, if the device is either heated or cooled depending on which side of the temperature spectrum it was exposed to, it will return to its normal operational behavior. However, if the part is exposed to temperatures outside of its normal operating range for too long, it will suffer permanent damage.

In the automation industry, temperature is often overlooked. It is quite common to see PLC stuffed into control panels with very poor ventilation and the machine itself stuffed into areas that will never get any ventilation. For example, machines that are too tall for buildings, such as cranes or even some industrial welding machines, will often be installed in holes in the ground. For machines that are too tall to fit into a given area, organizations will often dig out an area in the flooring to place the machine. Now, these are not normal holes that people dig out with a shovel; these are areas that are cut into the building, cemented over, and converted into workstations. These modified areas will allow organizations to fit the machine in the building; however, these areas will by nature often get little to no airflow. If you take into consideration that a normal manufacturing environment is usually not climate-controlled and that electronics generate heat, you can understand that temperatures inside of an electronic enclosure can easily reach triple digits, especially in the summer.

Outside of temperature, another factor that can cause issues with PLCs is debris.

Damaging debris

Another often-overlooked critical issue with PLCs is **debris**. Debris is often not even considered when troubleshooting a system. However, debris can be a very stealthy killer. Debris, especially the conductive kind, can easily kill a PLC if not properly accounted for. Due to the dirtiness of many PLC environments, corrosive and/or conductive material can find its way into the PLC. This problem is especially prevalent in PLC modules that have ventilation holes. In short, particles will find their way in through the holes and corrode the electronics or, if they are conductive, cause short circuits.

Debris can also lead to overheating. For modules that do require ventilation holes, dust, grime, and particles in the air can eventually clog the ventilation holes if left unchecked. This in turn will cause an excess of heat to build up in the unit, which will eventually lead to permanent damage to the module.

In all, most manufacturing environments often have copious amounts of debris floating in the air. This stems from the nature of manufacturing in general where there is often a dirty environment.

The final environmental factor that we're going to explore that easily has an adverse impact on the PLC is humidity.

The adverse impact of humidity on hardware

When designing or installing a machine, a silent killer that is overlooked is humidity. Humidity is especially dangerous in areas that are naturally moist and hot. Moisture is often a slow killer. The danger of moisture doesn't stem from the water droplets shorting out electronics (though they can); instead, moisture is often associated with corrosion. As moisture collects for long periods of time, it will often cause a buildup of rust and general corrosion that will eat away traces on a **printed circuit board (PCB)** and damage components on the PCB. In short, long-term moisture can have the same effects on a PLC as water does.

So far, we've explored a few common environmental issues that can have an adverse impact on PLCs. Though there are many more environmental issues that can adversely impact the performance of a PLC, we're going to explore a few common issues that are not necessarily environmental that can adversely affect it.

Understanding non-environmental issues

PLCs are machines just like your car, computer, or any other device. Much like those devices, there is a plethora of issues that can adversely affect your system. For PLCs, the first and most obvious problem is network issues.

Network issues

Whether you realize it or not, almost all PLC-based systems will use networks at some level. Even if your machine is not connected to a network, the PLC device will probably use some type of network communication protocol to communicate with peripheral devices such as HMI screens, wireless controllers, motor drives, and if applicable, other machines. Networking issues are among the most common issues that one will face in the day-to-day maintenance of a downed machine.

Network issues can come in many different forms and are often a root cause of the whole *broken software* fallacy. When network issues arise, you'll often see issues with devices such as motor drives not turning on, HMIs not displaying information, and a lack of communication between machines where applicable. Another common symptom is erroneous data that appears either on a PLC display, on the HMI, or on the computer controller itself. This data may be error codes or flat-out network warnings.

If networking is not the issue, another common problem is simple wear and tear.

Exploring wear and tear

It's no secret that most PLC systems are in constant use all day, every day. This means that parts wear out quickly and without warning. Though industrial components are designed to be much more rugged and have a longer lifespan than most off-the-shelf components, they will eventually wear out. It can often be hard to diagnose a failed component as the symptoms of a failed component can often be caused by other factors such as an environmental issue or other issues such as a poor network condition. Though it will be tempting to just replace a part that could be the culprit, you should only do so if absolutely necessary.

Another issue that may cause an issue with a machine's performance is calibration.

Exploring calibration

Calibration is vital for many machines. As parts wear in and get old or the general environment changes, machines will have to be adjusted to accommodate. In terms of unusual behavior, calibration is probably in the number-one slot. Most industries will require a machine to be routinely calibrated. The calibration interval can vary from industry to industry, machine to machine, and organization to organization. Many organizations will require at least a yearly and more often quarterly calibration. A good starting point in troubleshooting a machine is calibrating it. Generally speaking, a good calibration will clear up many erroneous behaviors.

So far, we have explored many issues not related to software that can cause erroneous behavior in a machine. Typically, if you hear an end user complaining of broken software, the error is typically one of the aforementioned. However, in extremely rare cases, software too can be the culprit.

Exploring erroneous software

Software will never just randomly stop working. If a program was working yesterday and the day before, and nothing has changed, the software is fine. However, there are times when software can be the culprit.

Software updates

Automation software is very dynamic. This means that, unlike traditional firmware, PLC software is often updated to accommodate for new processes and hardware. This can lead to erroneous behavior. Depending on the way the system is programmed, the update may wipe out the old calibration data or other preset values. There is also the possibility of bugs. When software is updated, a bug may have been accidentally introduced. This is especially true if the software is modified to accommodate a new hardware component.

One needs to be very cautious when updating software. Anytime you touch the source code of a machine, you must assume the machine is broken. This means when you're working on an update, you need to run a series of tests to ensure the machine still performs as expected. You also must allocate time after the initial test is conducted in case there are bugs in the system that are not immediately spotted.

Outside of software updates, another software-related issue can come in the form of corrupted software.

Exploring corrupted software

Source code will never change. A program typically lives in a file on the programmable device or is flashed to a chip. For software that is housed in a file, it is not unheard of for the file to become corrupted. Software that is flashed to a chip is not immune to this either and can sometimes become corrupted as well. Regardless, corrupted software will generally follow catastrophic events such as a sudden power outage, a hard power cycle, or an inexperienced programmer tampering with the wrong onboard systems such as kernel files.

In terms of behavior, all corrupted software will generally behave the same. It typically won't load. The corrupted software will, a vast majority of the time, prevent a program from properly loading. This is due to the nature of corruption. As we've explored throughout this book, software must adhere to a strict set of rules for it to compile and run. When corruption occurs, those rules will be turned effectively into mush, which will prevent the program from even loading.

All these issues that have been explored thus far are errors that will affect the PLC as a whole. These are common issues that can affect every module in the PLC except for the software errors, which will only affect the CPU modules. However, the PLC itself is not the only place where issues can stem from. In fact, it is often more prevalent to see faulty hardware than anything else. Therefore, in the next section, we're going to explore some common issues with hardware.

Common hardware issues

It is not usual to have to constantly change parts in a machine. Again, this stems back to the high usage of most PLC-based equipment, which causes wear and tear. In this section, we're going to explore some basic symptoms of faulty hardware. To begin this exploration, we're going to look at power issues.

Exploring power supply issues

As logic dictates, all PLCs and by extension, hardware in general, require a stable voltage of some kind. In the United States, most PLC-based equipment will draw power from a wall outlet or a high-voltage power drop of some kind. Typically, most PLC equipment will require a large power supply that is usually around 480 volts **alternating current** (**AC**). In other countries, this may vary depending on the local infrastructure, especially in places such as Europe where **direct current** (**DC**) power supplies are the norm. However, if you're in a place that utilizes AC power supplies, there will be a need for AC-to-DC converters that will drop the power down and convert the current type for digital devices, such as PLCs, that require lower DC voltages to operate. The way different AC-to-DC converters work will vary from device to device, but the general flow of a power supply will typically step down the voltage/current and then perform the conversion from AC to DC. The final output will be a steady supply of DC power.

As stated before, the way in which the power is converted will vary from converter to converter. However, as logic dictates, each converter is made of electrical components, which means they will eventually fail. As was covered in *Chapter 2*, most PLCs and industrial electronics in general require a stable 24 V DC to operate. When a power supply starts to fail, the voltage output will become unstable and start to either rise or dip. This will cause erroneous behavior with the PLC system. When a power supply has failed or begins to fail, you may notice some of the following symptoms:

- Intermittent reboots

- Random shutdowns and restarts

- Excess heat in the cabinet, more specifically near the power supply

- Smoke

- Over- or under-power messages from devices such as the PLC or supporting modules

- Completely dead system

- Blown fuses/breakers

- Random loss of data such as recently inputted job parameters

When symptoms such as these arise, you need to be very cautious with the machine. Not only could there be major safety issues such as the risk of fire or electrocution, but the PLC and other hardware could be put in danger too. Constantly power cycling (hard rebooting) a device can cause permanent damage to the electronics in the device. When the system constantly reboots itself, the software may become corrupted as well, especially if you're using a high-end PLC that uses an operating system such as Windows or Linux. Most higher-end PLCs will have onboard electronics that will prevent the device from dying in the event of a power loss; however, these are only temporary measures that are mostly used to protect the electronics. In many cases, the software can still be corrupted, especially if the problem persists. When symptoms of a bad power supply are noticed, it is best to shut the machine down and check the power supply out.

If you experience power issues, the power supply may not be the only culprit. If the device is drawing power from a wall drop, which most will, there is also a chance that the wall drop is bad. Having a bad wall drop or something wrong with the main power line, excluding something simple such as a blown breaker or fuse, is much rarer in most cases than a damaged power supply. If there is no power coming into the machine, the power supply will obviously not power up, which in turn means the rest of the machine is going to be powered down. In cases such as these, one can easily tell whether there is indeed a problem with the drop as other machine components, such as hydraulics, motors, and so on that do not rely on the power supply, will be dead as well.

> **Note**
> Wall voltage of any kind can be deadly. If you suspect that the problem is a wall drop, have a qualified electrician troubleshoot the issue!

Voltage may not be the issue you have. Other common issues you may experience will stem from the PLC itself.

Common PLC problems

In the previous section, we explored some basic issues that could affect a PLC-based system. However, what we have yet to explore in-depth are some of the behaviors that were alluded to. To begin the discussion, we're going to explore the effects of heat on a PLC unit.

Bad behavior due to heat

As we already explored, temperature can have a very adverse effect on a PLC's performance. A PLC will more often fail from overheating than from freezing. Typically, when the CPU module of a PLC starts to overheat, the PLC will start to behave abnormally. The way in which the PLC will fail varies, but systems that run Windows or Linux will typically freeze and then turn off. A key giveaway for any overheating problem is the system will freeze or fail to turn on for a period of time, usually about 10 to 30 minutes. Then the system will turn on again and, after a period of time, will repeat the same behavior.

The reason for the freezing and restarting stems from the temperature limits of the electronics. Depending on the system, the PLC may shut down due to being programmed to do so or the parts just reaching their operational limits. When the PLC is allowed to cool down, the PLC will restart and run normally again until it heats back up. As we will see later, this can be an easy problem to diagnose and fix.

Temperature may not be the only problem that can lead to unexpected shutdowns. Sometimes, if your PLC is shutting down and you know it's not your power supply, you may have a bad battery.

Bad batteries

Believe it or not, a common problem that can cause issues with PLCs is simple batteries. Many CPU modules, such as ones manufactured by Beckhoff, will often use small CMOS batteries to keep track of times and dates among other things. The batteries will usually last years and the issues that stem from the bad battery may vary depending on the device, manufacturer, and so on. Common behaviors of a bad battery will be akin to wrong dates/times, error messages, beeping noises, PLC shutdowns, and of course, warning lights coming on or blinking. Therefore, if you see unusual lights blinking, hear beeping sounds, experience shutdowns, or have erroneous dates and times, you could have a bad battery. Typically this is a CR2032 CMOS battery. Due to the size and shape of the battery they are commonly referred to as watch batteries.

Outside the issues mentioned previously, a catastrophic issue that can pop up is often a very unexpected one. A major problem that is often associated with bad batteries comes into play when the battery is installed backward, or the insulator tape isn't taken off. Depending on the make or model of the device, if the tape is left on or the battery is installed backward, the PLC won't turn on. Outside of not turning on, a battery that is installed backward can lead to device damage! So, be very careful when it comes to battery maintenance.

When it comes to maintenance, a very common issue that we explored earlier is communication failure between devices. There are an infinite number of root problems that can cause these issues; however, the symptoms will usually be the same: erroneous communication between devices. In the upcoming section, we're going to explore some common hardware symptoms that stem from networking issues.

Bad behavior from networks

All digital devices have addresses. An address to a digital device is like a home address to a house, like the way a person can send mail to another person by using their address. Digital devices will send data to each other using their unique addresses. With the widespread adoption of wireless devices such as wireless controllers and smart devices, networking issues are becoming more prevalent. As we explored before, a common hiccup in many modern PLC systems is networking. The root cause of many issues stems from the hardware having the wrong addresses.

There are many different types of addresses that are used in networks and, more specifically, automation. A common type of address is the classic IP address. An IPv4 address will look like the following: `192.168.10.99`

For an IP address to properly send data through the network, the address must typically share the first three numbers while the fourth must be unique. For example, if you wanted two devices to communicate, they would need to have the following address: `192.168.10.xxx`

One device could have the last digits as 10 while a second could have the last digits as 999.

Another issue that might pop up is whether there are two devices with the same address. Many systems will safeguard against this, but it can still happen. In short, having two devices with the same address is like having two houses with the same address. When the post person tries to deliver the mail, they may not know which mailbox to put the letters in. In the case of the digital system, when a data packet is sent, the system may get confused and not know which device should receive it!

Typically, if two devices are no longer communicating or are getting unexpected data communications, especially after a hard reboot, there could be a problem with the systems' addressing configuration. In terms of automation, there are many ways to set an address for a device. Some addresses are programmed into the device and others are set manually with switches, while other, more sophisticated devices use traditional IP addresses. No matter what mechanism is used, if they are being set manually, it is very common and easy to introduce issues by mistake. So, if you're having issues with devices communicating with each other, the first step should be checking communication between the devices.

Outside of addressing, another major issue that can arise is with network cabling.

Network cabling issues

Outside of addressing, another major issue that is often associated with bad networking is cabling. Even in today's wireless-dominated world, network cables of all kinds are still widely used, especially in automation. Due to the movement and high-maintenance nature of most PLC-based machines, it is very common to break cabling. Depending on how the machine is set up, it is very common to pinch the cable and sever its internal connections; that means it's common to break the wires in the cable. This is mostly due to the moveable nature of most PLC-based machines. For systems such as robots or other devices that move, it's common to need snake communication cables such as drive cables in the machine. Unfortunately, as machines are poorly maintained or as they simply age, the cables can get pinched and break.

Outside of pinching the cable, simply unplugging and plugging the cable can lead to damage. If it is not properly removed or inserted into its port, pins can be bent, wires can be broken internally when the technician pulls on them, and of course, the plastics that secure certain cables in the port can break. When any of these failures are experienced, communication can become unstable at best. There may be intermittent drops or flat-out communication failures.

Devices that usually communicate via network cabling are as follows:

- HMI screens
- Motor drives
- Robots
- Certain sensors
- General network equipment

Pretty much anything that transmits data can use a network cable of some kind.

Though it goes well beyond the scope of this book, many networks have physical hardware components such as routers and switches. This is especially true for systems that are integrated across a plant or factory. Much like any other physical hardware, these parts can fail. Therefore, if you encounter networking issues, and the cables are good and the addressing is set right, there could very easily be a problem with a switch, router, or other routing component.

So far, we have explored some common causes of issues as well as some of the symptoms and hardware components that could cause the issues. However, as we have seen, certain symptoms can be caused by different things. In the next section, we're going to look at some techniques to troubleshoot and pinpoint the problem.

Exploring troubleshooting techniques

We now know what some problems that could affect the performance of a PLC are, and their common causes. As anyone who has ever worked on an electromechanical system knows, simply knowing what a cause could be is not the same as knowing what the cause is. In this section, we're going to explore some troubleshooting techniques to help pinpoint a problem.

Before we get into troubleshooting, we need to look at the tools that every engineer should have in their toolkit.

The PLC toolkit

There is nothing more embarrassing than showing up to a service call or being asked to troubleshoot a PLC and not having the necessary tools. The following is a list of common tools that everyone who is tasked with troubleshooting a PLC-based machine should have:

- **Screwdrivers**: Everyone tasked with fixing a machine should, at the minimum, have a series of flathead and Phillips screwdrivers of varying sizes. Usually, it's a good idea to get at least two of each size. Generally, a better idea is to pick up two packs of assorted screwdrivers of both types.

- **Pliers**: A technician or engineer should have at least a pair of needle noses, channel locks, wire cutters, and other assorted pliers. Usually, one pair of each will work but ideally, you would want a couple of pairs of each.

- **Wire strippers**: Often, a tech or engineer will have to rewire things, and a decent pair of wire strippers will be needed to strip the ends off the wires.

- **Wire crimps**: These are plier-like devices that can be used to attach solderless connectors to the ends of wires.

- **Thermal gun**: An often-overlooked tool that many techs or engineers will not always have on hand is a thermal gun. This tool is vital for troubleshooting temperature issues.

- **Multimeter**: Every tech/engineer should always have at least one quality multimeter such as a Fluke on them at all times. Typically, a pair of alligator clips, probes, and fuses should also be included. It is also wise to include several fresh 9V batteries for the meter as well.

- **Laptop computer**: The toolkit should always include a laptop computer with the necessary programming software and source code loaded onto it.

- **Network cable**: It is wise to include a programming cable to interface with the PLC. One should also include multiple cables such as networking cables that are used in the system.

- **Flashlights**: Every kit should include a magnetic flashlight that can stick to the wall of the cabinet. It is also wise to have batteries on hand for the flashlight. A tech/engineer should have at least one flashlight; however, two are ideal.

- **Brushes**: It is usually a good idea to have paintbrushes of varying sizes on hand. These brushes can come in handy when you need to clean out ventilation ports or other small crevasses.

Now that we have a toolkit, let's start troubleshooting by looking at possible power supply issues.

Diagnosing power supply issues

One of the easiest issues to troubleshoot is a power supply issue. The power supply is usually an obvious problem. Typically, when the power supply fails, the system will not turn on. However, a dead machine does not necessarily mean a bad power supply. There could be a problem with a breaker, a fuse, a switch, or even the wall power drop.

The first step in troubleshooting is to simply check the wall breaker. Believe it or not, it is very common for someone to flip a breaker and forget to turn it back on. Before you even touch a tool, you want to check to ensure all the breakers and wall switches are on. If the wall breaker is on, you want to move down the line and check inside the cabinet. Typically, machines will have either a main fuse(s) or breaker(s). The breakers will be easy enough to check. Checking a fuse can be a little more in-depth but it is still easy to do. All you need to do is use the diode setting on your multimeter to check the fuse. If the meter beeps, you typically have a good fuse.

If all is good, the next thing to do is check the power supply. To do this, check the voltage at the DC output with your meter. Typically, power supplies will output 24V; however, that value may vary from supply to supply so be sure to check what the power supply output should be. Regardless, if you have the correct voltage on the power supply output, move down the line again and check at the end of the wire(s) leaving the power supply. If the output is good in the power supply but dead at the end of the wire, this is typically indicative of a bad connection, and more than likely, a bad connector at the wire end(s). If you do have stable voltage at the end of the wires, the problem is more than likely with the electronics.

If you are getting sporadic booting issues or have a dead PLC, you can try checking the temperature.

Diagnosing temperature issues

If the PLC is dead and you do have power going into the PLC, the next thing to check is the temperature. Depending on the PLC, there may be built-in warnings; however, you may not always have that luxury. If the PLC is constantly rebooting or shutting down sporadically, it may be a temperature problem as explored previously. To confirm, turn the device on and let it run for a little while. While the PLC is running, try to use a thermal gun to see whether the PLC or any other component is heating up.

If the device is heating up, there could be multiple reasons, as were already explored. It could be due to the ventilation holes becoming clogged with debris, metallic debris causing shorts in the devices, a failed electrical component, or simply the heat in the cabinet building up. In terms of troubleshooting this problem, the first thing you can do is try cleaning out the ventilation ports. You can typically do this with a small brush or paper towel. If this isn't enough, there could be something in the device itself causing a short. This is a much harder problem to fix as the debris will be hard to remove. However, you can get lucky by trying to blow out the device with compressed air. If the problem persists, there is a good chance an electrical part has failed. If a part has failed, the module will need to be repaired or replaced. Now, if a module has failed, they are usually not worth repairing. It is usually cheaper and easier to simply swap the part out.

Moving on to the next topic, we're going to look at troubleshooting network connections!

Diagnosing networks

Troubleshooting networks can be a daunting task if you're not familiar with how network devices work or are not experienced with IT in general. PLC programming software usually has a built-in way of detecting devices. This means if you're working with a device of those types, you can simply use built-in tools to troubleshoot device communication. If your device is a typical piece of IT equipment or a Windows or Linux PLC, you can also connect to the device or network and ping it using the ping utility. The way you ping a device is to first give your PC a compatible IP address and connect to the network. Once you connect to the network, you should open the command prompt on a Windows or Linux terminal and run the following command:

```
ping xxx.xxx.xxx.xxx
```

Here, the x represents a digit in the IP address.

If you get a response, the network connection is working as expected; if you do not get a response, there is an issue. This will work at a high level with devices such as the PLC, network devices, and certain other hardware. When it comes to devices such as modules, motor drives, and so on, pinging may not work and you may need to depend on the device software to troubleshoot.

If you suspect there is a problem with the network communication, there could be any number of problems. However, some common fixes are as follows:

- Change the communication cables
- Power cycle the device
- Check whether there is a firewall that is preventing communication
- Temporarily change out a suspected defective device to see whether it fixes the problem

Network issues can either be hardware or software-related; however, if the system has been up and running for a while, chances are something went wrong with a piece of hardware. For many existing applications, a problem with networking will usually boil down to broken cables, hardware that needs to be power cycled, or hardware that needs to be replaced.

There are many other issues that can cripple a machine or series of machines. Though it is very rare, software can still pose an issue if it is modified or newly installed. As such, before we move on to the final project, we're going to look at some basic software troubleshooting techniques.

Troubleshooting software

If a problem is software-related, it is usually because the software was recently installed (new machine) or modified. Troubleshooting can be a very complicated task and takes experience and skills to accurately troubleshoot. This section is going to give a quick crash course in some basic troubleshooting techniques, starting with learning how to back up software.

> **Note**
> It is never a good idea to modify the software to work around a broken or malfunctioning part.

Backups

Before you even think about touching the software on any machine, you need to have a backup of the software handy. If PLC systems support them, there are technologies such as Acronis that can clone and restore an entire system. Many of these technologies will create a snapshot of the current system along with its configuration. If you have a PLC or system that can support the technologies, it is worth the money to invest in it. If your system does not support these technologies or you don't have access to them, you can still archive the software. Almost all PLCs have the capability to download the code that is currently burned on the PLC. Usually, the only software that is hard to download from the source is the HMI. However, the HMI should be as "dumb" as possible and you typically won't need to archive it.

In terms of troubleshooting, the very first thing you should do is either make a clone of the system, using a technology such as Acronis, or download the current PLC code along with any other piece of software that you can preserve. This may seem counterintuitive; however, you want these backups as a restore point if the problem cannot be resolved or the problem is accidentally made worse. Once you have fixed or modified the code base, you need to save it somewhere.

Introducing version control

There is nothing quite as bad as arriving at a service call and being asked to fix a machine's software and having the wrong code/backup, or having to flat-out tell a customer that their machine's code is simply lost. This is where version control can come in handy. Version control is very important, but many automation companies do not see the need for it, nor are they willing to invest in or use it. Not using version control can cost a company copious amounts of money by causing unnecessary rewrites for lost software. Version control is a software tool that can version software. This means that when a new version of the program is committed, the changes are tracked. If you need to use an older version of the program, the changes from the subsequent commits can be ignored and you have the needed version of the program. So, if you made a change to support Motor Drive A and you need the original program that supported Motor Drive B, you can simply pull that version of the software. Source control also allows an organization to store archives such as Acronis backups, wiring diagrams, and any other type of machine documentation.

There are many different version control tools out there. One can use the following:

- GitLab
- GitHub
- TortoiseSVN
- TFS
- BitBucket

Many others can also be used. A great example of a version or source control system is GitHub, which stores all the code for this book. Most of these will either use Git or SVN. SVN is much older, and Git is by far more popular. In all, adopting one of these systems will provide the following benefits:

- Prevent the loss of older iterations of the program
- Prevent code from being lost
- Keep code bases in a single location
- Allow for sharing and collaboration of a given project
- Allow you to store documentation and archives

Though it will cost a small amount of money to use a version control system and it will require a bit of training, it is well worth it in the long run as it'll save you money.

Now that we know how to archive and store software, let's look at troubleshooting software.

The basics of troubleshooting software

Learning how to troubleshoot software is as much an art as it is a science. This section is going to explore the basics of finding or troubleshooting a problem in code. The first troubleshooting technique we're going to look at is print debugging.

Print debugging

Print debugging is a troubleshooting technique that allows you to view the various locations of the program. For example, suppose we want a program to display a message when a number is greater than 23. To do this, we could use the following code:

```
PROGRAM PLC_PRG
VAR
     debugMsg  : STRING[20];
     outputMsg : STRING[15];
     number    : INT := 21;
END_VAR
```

In this example, we have a `debugMsg variable` that will have 20 characters, `outputMsg` that will hold a message, and finally, a `number` variable for the program to test against. The logic of the program will be as follows:

```
IF number > 23 THEN
     outputMsg := 'number is greater than 23';
END_IF
```

If you run the program, you will notice that the program does not produce the message as expected. As such, we can put some debug messages in to help see where the program is failing. We can modify the code to match the following:

```
debugMsg := 'start of program';
IF number > 23 THEN
     debugMsg := 'In the if';
     outputMsg := 'number is greater than 23';
END_IF
```

Here, we have a `debugMsg` at the top, and since the message we want is in the `IF` statement, we can drop one in there too. When we run the program, we should get the result shown in *Figure 16.1*. Notice that the debugMsg value is set to `start of program`:

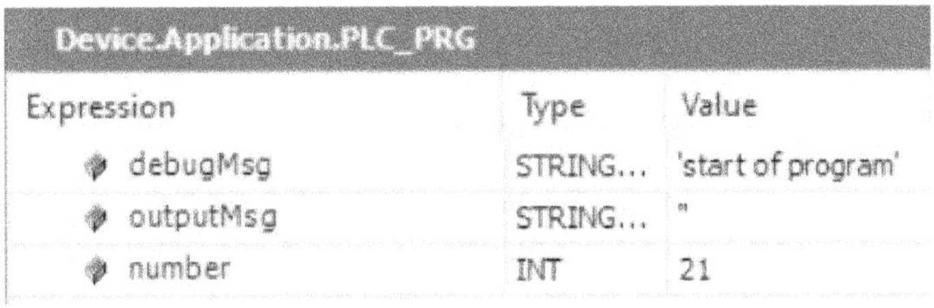

Device.Application.PLC_PRG		
Expression	Type	Value
🔶 debugMsg	STRING...	'start of program'
🔶 outputMsg	STRING...	"
🔶 number	INT	21

Figure 16.1 – Debugging output

This output means that for some reason, the code in the IF statement is not running. If we examine the code and consider the original use case for the program, we can see that this program is designed to test whether a number is greater than 23, and our number is set to 21. This looks like a mistake, and we can check it by writing a new value.

Exploring writing values as a troubleshooting technique

So far in the book, we've used writing to change values; in a sense, we've also used writing to explore code behavior. We can use this same technique to troubleshoot as well. This is a very common practice for PLCs that can support dynamic writing.

> **Note**
>
> Writing a value and forcing a value are similar in nature but operate differently. A forced variable is permanent and must be changed by the user, while writing a value can be easily overwritten. Forcing a value can be very dangerous and the value must be unforced by the user!

For this example, we can write a value to troubleshoot. Since we have the number set to 21 and our program is testing for a number that is greater than 23, we can write a value of 24 to see what happens:

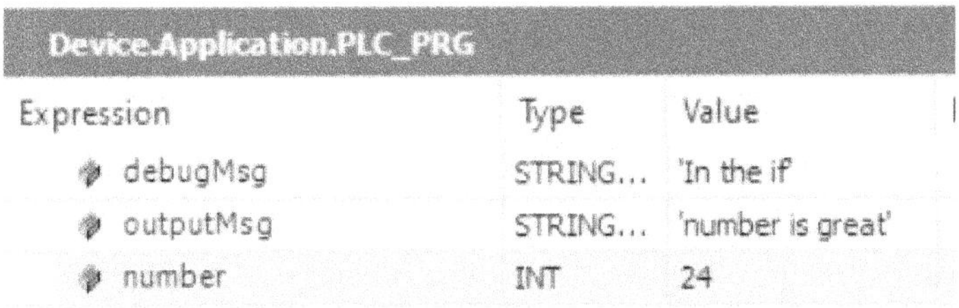

Device.Application.PLC_PRG			
Expression	Type	Value	
🔶 debugMsg	STRING...	'In the if'	
🔶 outputMsg	STRING...	'number is great'	
🔶 number	INT	24	

Figure 16.2 – Writing value output

As we can see, this time, when we wrote 24 to the number variable, the program worked as expected. We got the message we put in the IF statement, and we got the message that we originally expected!

Now that we've explored software troubleshooting, we should have a decent grasp of the basics of troubleshooting. This means that we can move on to the final project.

Final project

For the final project, let's assume the following scenario.

You were recently contacted by a customer about a machine with broken software. The machine will boot up and run as expected for about 25 minutes, then unexpectedly shut down. The customer says the issues started about two days ago and is sure that the software broke unexpectedly.

With that, let's look at how we can troubleshoot this.

Troubleshooting

First, since the machine has been running as expected for at least 25 minutes and the system has just started behaving this way, we can assume that the software is not broken. There could technically be an issue with the PLC's operating software such as the operating system, or the user is doing something that the machine wasn't intended for and programmed it to do something out of its bounds. However, that whole scenario is unlikely. Based on what we learned, this sounds like there is an issue with the PLC overheating or with the power supply.

When we get to the site, the first thing we notice is that the machine was in a pit – that is, in a carved-out hole in the facility to accommodate the height of the machine. When walking into the facility, we notice that it is hot and humid. Since the device is powering up, there is probably nothing wrong with the breakers, fuses, or wall power.

Challenge 1

Try to think of a few possible issues that could arise from the environment.

Solution

Right off the bat, the power cycles, humidity, and heat could mean that the system is overheating.

Troubleshooting steps

The following steps can be used to for troubleshooting:

1. Inspect the machine in the off state. Look for water droplets and feel whether the inside of the cabinet is hotter than usual. Also, visually inspect the ventilation ports on the devices.
2. Turn the machine on and let it run. When the device cuts off, quickly open the cabinet and use the thermal gun to check the temperature.

Results

The device reads about 90 °F. This is toward the high end of the spectrum but still within range. So, we're going to clean off the ventilation ports and leave the cabinet door open to vent any excess heat.

Upon doing this, the machine still shuts down after about 25 to 30 minutes. So, we can move on to the next possible issue: the power supply.

To troubleshoot the power supply, we turn on the machine.

Challenge 2 – troubleshoot the power supply

Think out the steps needed to check the power supply.

Solution

Power on the machine and use your multimeter to measure the power going into the PLC. After about 15 minutes of running the power supply, you notice that the output voltage is starting to drop. You also notice that the power supply is starting to heat up.

Troubleshooting steps

1. Keep measuring your voltage output.
2. Use your thermal gun to measure the temperature of the power supply.
3. Visually inspect the ventilation ports on the power supply.

Results

You notice that as the temperature rises on the power supply, the voltage begins to drop. This means that the power supply could be overheating and failing for that reason. So, you try cleaning out the power supply ventilation ports. After running the machine, it stays on this time without shutting down. The problem is solved!

In this case, debris that built up in the power supply ventilation ports was causing excess heat buildup and causing it to fail. At this point, you can either opt to call it a day or swap out the supply. It would be a best practice to change out the supply, but some customers may not want to pay for the part if the old one is working.

Summary

This chapter covered the basics of troubleshooting. We explored hardware issues including common problems such as heat and common component failures. We also explored software-related issues such as bad source code and even networking issues. There is no magic bullet to troubleshooting a problem. What you'll find is that no two problems will ever be the same. Nonetheless, this chapter has provided some basic techniques and issues that could be used in your day-to-day life!

Wouldn't it be nice if we could ask our computer what's going on and it gives us a list of possible solutions? Well, that dream can be somewhat of a reality with generative AI. In the next chapter, we're going to explore how generative AI (ChatGPT) can be used to help us.

Questions

1. What is a symptom of a bad power supply?

2. What are a few tools you should have in your toolkit?

3. What does the ping command do?

4. What is a symptom of a network issue?

5. What is a possible symptom of a failed CMOS battery?

6. What is version control?

7. Name two benefits of version control.

8. What is print debugging?

9. What is the difference between forcing and writing a value?

10. How should you respond to broken software?

17
Leveraging Artificial Intelligence (AI)

Over the past two years, two little letters have taken the world by storm. Those letters are **Artificial Intelligence (AI)**. The last two years have seen a revolution in the way everyone, from students to content creators, are going about their daily lives. Software engineers and, by extension, automation programmers, are no different. Systems such as Copilot, Gemini, Devin, and, of course, the all-too-famous ChatGPT have captivated the attention and imagination of developers, and many are desperately trying to integrate these systems into their daily work lives.

The world is abuzz with how developers are trying to integrate **generative AI (GenAI)** into their daily workloads. Due to the novelty of the technology, many are still struggling to figure out just what GenAI can and cannot do. There are many articles saying how it can be used as an assistant for programmers, while others say that GenAI will render human programmers obsolete. This chapter is going to look at how to use generative technology such as ChatGPT and dispel some misinformation along the way. To do so, we're going to look at the following topics:

- What is GenAI?
- What GenAI can't do
- Reasonable expectations with GenAI
- The basics of prompt engineering
- Producing workable code with ChatGPT

To round out the chapter, we're going create some prompts that will generate an `IF` block for a temperature range program that we can use in a project!

Technical requirements

This chapter will use ChatGPT to generate our code for us. This means we're not going to write any code. The code generated by ChatGPT can be downloaded at the following URL:

`https://github.com/PacktPublishing/PLCs-for-Beginners`

To follow along with this chapter, you will also need a free account for ChatGPT. For this chapter, we're going to use the free version, 3.5 at the time of writing this book, to experiment with. You can sign up for free at the following link:

`https://openai.com/`

What is GenAI?

The term *AI* is obnoxiously overused in today's world. AI is used by laypersons to refer to almost anything. Up until recently, anything that was related to automation was considered AI. However, the terms have gotten a little more accurate as of late, and true AI is starting to be distinguishable from general automation. So, the first topic that we need to explore is what AI is!

What is AI?

The most accurate definition of AI is a program that can perform a task without being specifically programmed to do so. This means that though PLC-based systems are sometimes referred to as AI, they really don't meet the qualifications of being AI. Automation systems, such as the theoretical ones explored in this book, must be programmed to do a certain task. For example, if you recall back to our color and shape sorter, we had to specifically program it to sort parts based on their shape and color. A true AI would be trained using data and a model to recognize things such as stars and colors and sort them accordingly. AI can be integrated with PLC-based systems, but by default, they are not AI systems. An AI system will require two vital components to work: one is called a **model**, and the other is called **data**. To begin our exploration into AI, we're first going to explore what training data is!

What is training data?

Training data is used to teach an AI system how to do something. Data comes in many shapes and forms; for example, a system can use **labeled data** or **unlabeled data**. Labeled data is data that is preprocessed by a human. That is, if the dataset consists of various pictures of fruit, a human will need to go in and mark what's an apple and what's an orange. Unlabeled data does not have any markers and is not processed in any way. The system will process the data using what is called a model to try to find patterns within the data. This means the next step is understanding what a model is.

What is a machine learning/AI model?

As was touched upon in *Chapter 1*, a **machine learning (ML) model** is an algorithm that is used to process data. As we previously discussed, there are many different types of models, and many of the types have subtypes. Regardless of the model type, it can be supervised, unsupervised, or a hybrid of the two. A supervised model will use labeled training data, while an unsupervised model will use unlabeled data. Finally, a hybrid model will use a combination of both. The next logical step in understanding what a GenAI model is to explore what a **large language model (LLM)** is.

Understanding LLMs

An LLM is simply an AI model that is specialized in understanding and producing human language. The best way to think of an LLM is as a chatbot on steroids. These models are trained on very large datasets and excel at things such as language generation, text completion, language translation, and more. More specifically, LLMs are designed for **natural language processing (NLP)**, where NLP is an AI system in which humans can use natural speech to interact. That is, an NLP system, unlike a PLC, does not use a specific programming language for interactions. The NLP is really what makes GenAI what it is.

What is GenAI?

GenAI is a type of AI that can be used to generate content. That is, GenAI is used to create text, music, images, and even programs among other things. For our purposes, we're going to use AI to help write PLC code. Systems such as ChatGPT will draw on very large training data to do any number of programming tasks that go well beyond simply writing code. These generative systems can be used for any number of tasks, from writing code to providing troubleshooting feedback. In all, when used properly, these generative models can produce fairly high-quality code.

At the time of writing this book, anyone who has been paying attention to the news may notice that many companies are experimenting heavily with these AI systems. There is a lot of talk about what these systems can do and how they can replace human programmers. With all that, the next section is going to explore what these generative systems can't or shouldn't do.

What GenAI can't do

At the time of writing this book, there is a bunch of hoopla about companies trying desperately to leverage AI to replace their human workforce. As of late, there has been a lot of talk of AI replacing human workers, and this is further compounded by many news stories of large companies laying off thousands of workers. So, a logical question is, what does this mean for us programmers?

Much as with any new technology, it is drastically overrated. Though it is still early in its development, this infant technology cannot fully replace human software developers. Though it can produce quality code and help developers understand and diagnose issues, we will see later on that stock AI systems such as ChatGPT 3.5 produce quality, but compared to the hype, underwhelming code. In all fairness, this is with the standard, non-optimized version of the AI systems. As time progresses and the technology matures and becomes optimized, this may change. However, for now, a company that tries to replace all its programmers with GenAI is a company that is trying to doom itself.

As of right now, the technology is very immature. As stated before, this may change, but right now, the limit of a stock GenAI's ability is limited to what is on the internet. This means if you're facing a novel problem or need a novel solution to an existing problem, GenAI probably isn't going to help much. The best you're going to get out of these stock GenAI systems is a point in the right direction. Unfortunately for those hoping to replace their human development team soon, the technology simply isn't there yet.

Stock generative systems are currently not very good with the following:

- Analyzing complex problems
- Producing high-quality code bases
- Carrying out complex tasks
- Replacing human developers

This is mostly true for stock platforms such as ChatGPT; however, as stated before, there are some optimized systems that are used for software development. These systems claim to be able to analyze and produce complex software. They are also very new, and whether they can actually replace a human programmer or not is still up for debate. If GenAI can't replace a human development team, what can one expect when utilizing a GenAI system?

Reasonable expectations with GenAI

It is true that GenAI is changing the landscape, especially for entry-level software developers and, by extension, entry-level PLC programmers. It is arguable that the true benefits of GenAI have been vastly blown out of proportion. This book utilized ChatGPT to generate some basic example code, mainly with sorting, and the results were marginal at best. The code worked, and it did look like a fairly experienced programmer wrote it in terms of structure and efficiency, but it did take some know-how to implement it. When it comes to GenAI, we first need to look at what not to expect.

What not to expect with GenAI

As touched upon before, all you potential hiring managers looking to replace your dev team with ChatGPT are going to be disappointed. ChatGPT and other similar systems are not suitable replacements for human programmers. The jury is also still out as to whether the GenAI systems that are designed to program are going to be able to replace developers anytime soon. The biggest hang-up to GenAI systems is producing solutions for novel problems. That is, if you have a unique problem that is outside of the model's training data, the model is at best going to give you a general direction of how to proceed. Most of these models were trained on data that was collected from the internet. This means that if the problem is unique to your code base, the AI is going to have a tough time figuring out how to solve it.

Another area GenAI is going to let you down is with implementing code. Quite often, systems such as ChatGPT will go off into the weeds with generated code. Depending on how you write your query, aka **prompt**, the generated code may be nearly useless. There are also issues with deploying the code. One aspect of ChatGPT that a lot of people overlook is that, at best, it's only going to give you code. Unless you're using an optimized GenAI that can interface with your respective PLC, you will still need a human to go in and physically upload the code to the device. Sure, there are workarounds for this, such as using a GenAI **application programming interface** (**API**) and developing a system that can do that for you or using a specialized system; however, this will cost more time and money in the long run and still probably won't give you the results you're looking for.

What can a GenAI be used for? If it can't replace a human programmer or give us usable code 100% of the time, what good is it? Why should we even consider using it? This is where reasonable expectations come into play. In the next section, we're going to look at some use cases as to how GenAI can help!

What to expect when using GenAI

The best way to think of AI is to think of it as an assistant. The fallacy that is going to sink many companies is that they are trying to think of GenAI as a programmer. That is, companies are desperately trying to replace carpenters with automatic screwdrivers. A better way to think of GenAI is as a tool that can be used to help programmers be more productive.

One way that GenAI can be used to help developers is to have it write standard code. For example, if a user needs to implement a sorting algorithm but may not remember the code off the top of their head or just wants a quick and easy solution, a system such as ChatGPT can come in handy.

Another use for a GenAI is to help troubleshoot small blocks of code. In cases such as this where maybe a value is being computed wrong or a piece of code is crashing, systems such as ChatGPT can be used to help pinpoint the problem.

GenAI systems can also be used as research assistants. In automation programming, things aren't always as well documented as their traditional programming counterparts are. This means that it can take an exorbitant amount of time to find relatively simple solutions to common problems. For tasks such as these, GenAI can come in handy. Due to the NLP nature of these systems, needing an answer to a common problem is as simple as asking a question!

To summarize, GenAI is best suited for tasks such as the following:

- Answering common questions

- Generating simple solutions for common problems

- Helping a developer troubleshoot a problem

- Helping a developer solve common problems

On the other hand, GenAI is not so well suited for the following:

- Replacing developers

- Finding solutions to complex or novel problems

- Single-handedly writing large and complex code bases

- Deploying solutions

The key takeaway here is that **GenAI is a tool**. Though this is a rapidly growing and changing field, as of writing this book, GenAI cannot and should not replace programmers. As generative technology progresses, this may change in a few years, but as of now, anyone who tries to use GenAI as anything other than a tool, at least in the development world, is doomed to failure.

The next step in understanding GenAI is understanding how to interact with it. On the surface, using a system such as ChatGPT is straightforward and not that different from using a search engine. In the next section, we're going to explore the basics of interacting with ChatGPT!

The basics of prompt engineering

For this section, you will need to set up an account to use ChatGPT. You will have the choice between a free account that uses GPT 3.5 or a paid account that uses 4.0. For this book, we're going to use the free account. At any point, if you want to, you can upgrade to the paid account! Once you're set up, you will be free to move on to creating your first prompt!

Creating a prompt

The key to interacting with ChatGPT or any other GenAI is understanding what a prompt is. In my earlier writings about ChatGPT, for better or worse, I almost always exclusively referred to a prompt as a query. I did that because, in my opinion, the best way to interact with GPT is as a tool that can answer questions. These questions may command the system to generate a code block, email, or simply answer a general question, and though they may be worded as commands to ChatGPT, they have their roots in questions. In my opinion, a great prompt has its mental basis in the form of a query. That is, you need to know something, and you're asking ChatGPT to answer it.

Our first prompt

There is no better place to start learning how to write prompts than by asking ChatGPT what its suggestions are. So, for this tutorial, log in to ChatGPT and enter the following:

```
What are the basics of prompt engineering?
```

Run the prompt, and ChatGPT will output something akin to the following:

```
Prompt engineering involves crafting effective prompts or instructions
to direct a language model's responses towards desired outcomes. Here
are the basics:
1. **Understanding the Goal**: Define the objective or desired outcome
of the prompt. This could be generating creative content, answering
questions, completing tasks, or any other specific goal.
2. **Analyze Examples**: Review existing examples to understand how
the language model responds to different prompts. This helps in
identifying patterns and formulating effective prompts.
3. **Keyword Selection**: Choose keywords or phrases that are likely
to elicit the desired response from the model. These keywords should
be relevant to the task at hand and provide clear direction.
4. **Clarity and Simplicity**: Keep prompts clear, concise, and easy
to understand. Avoid ambiguity or complex language that could confuse
the model.
5. **Provide Context**: Contextualize the prompt if necessary to
provide the model with additional information or background to
generate accurate responses.
6. **Experimentation**: Test different variations of prompts to see
which ones yield the best results. This involves tweaking keywords,
adjusting phrasing, and exploring different formats.
7. **Iterative Process**: Prompt engineering is an iterative process.
Continuously evaluate the model's responses and refine the prompts
accordingly to improve performance over time.
8. **Consider Ethics**: Be mindful of the ethical implications of the
prompts you create. Avoid biases, stereotypes, or prompts that may
generate harmful or inappropriate content.
9. **Documentation**: Keep track of successful prompts and their
corresponding outcomes. Documenting your experiments and findings can
help inform future prompt engineering efforts.
10. **Community Engagement**: Engage with the prompt engineering
community to share insights, exchange ideas, and learn from others'
experiences. Collaboration can lead to new approaches and best
practices in prompt design.
```

The prompt response may vary. ChatGPT will sometimes give radically different answers even if you input the exact same prompt.. Therefore, your response may look very different especially when it comes to the code it generates. It is important to remember that the response you receive will be the same at heart but may and probably will differ. As can be seen, prompt engineering is a skill, and it's an iterative process. The best way to interface with ChatGPT is to approach it the same way you would have a conversation with a living person.

The key to prompt engineering is detail; you want to add sufficient detail the same way you would with a living person. The more information you can give the system, the better your responses are going to be. For example, consider the following prompt:

```
What are the basics of prompt engineering, limit the response to the
five most important.
```

This prompt asks ChatGPT to give the basics of prompt engineering, but we added the extra detail of only wanting the top five most important features of prompt engineering. The response to the prompt is the following:

```
Certainly! Here are the five most important basics of prompt
engineering:
1. **Clear Objective**: Clearly define the goal or desired outcome of
the prompt to guide the model's response effectively.
2. **Keyword Selection**: Choose relevant keywords or phrases that
direct the model towards the desired task or topic.
3. **Clarity and Simplicity**: Craft prompts that are easy to
understand and free from ambiguity to ensure accurate responses from
the model.
4. **Contextualization**: Provide necessary context or background
information if required to help the model generate appropriate
responses.
5. **Iterative Improvement**: Continuously refine and adjust prompts
based on the model's responses, experimenting with variations to
optimize performance over time.
```

This prompt produced a similar result; however, the key points were a little more targeted.

Learning how to produce quality prompts is a lot like producing quality Google searches. When you're first learning how to effectively google things, you stumble around until you find your own unique way to query the search engine. The same can be said with prompt engineering. It'll take some time and a little practice to get used to using the system, but once you do, you'll have a very powerful tool at your disposal.

The same can be said for engineering prompts that produce workable software. In the next section, we're going to explore writing a few prompts that will produce workable code that we can plug into CODESYS.

Producing workable code with ChatGPT

As stated before, code from ChatGPT may need a little massaging for it to compile and run. A part of this stems from the way prompts are written, while some of it is due to the way the system was trained. To see the type of code that ChatGPT produces, let's ask it to create a program that can count from 1 to 10. To do this we're going to use the following prompt:

```
Write an IEC 61131-3 Structured Text program that can count from 1 to
10.
```

After running the prompt, ChatGPT is going to return something like the following:

```
PROGRAM CountingFrom1To10
VAR
    count : INT := 1; (* Counter variable *)
BEGIN
    WHILE count <= 10 DO
        (* Output the current count *)
        (* Replace this line with your output method depending on your
PLC *)
        // For example: Display(count);

        (* Increment the count *)
        count := count + 1;
    END_WHILE;
END_PROGRAM;
```

As we can see right off the bat, the generated script will produce code that cannot be copied and pasted into CODESYS. We have some odd keywords that are not necessary, such as the END_PROGRAM and BEGIN keywords. If you copy and paste in this code, you're going to get something akin to the output seen in *Figure 17.1*:

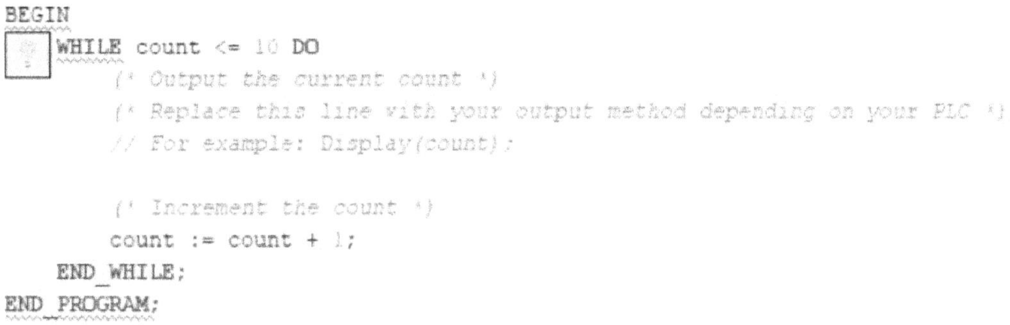

```
BEGIN
    WHILE count <= 10 DO
        (* Output the current count *)
        (* Replace this line with your output method depending on your PLC *)
        // For example: Display(count);

        (* Increment the count *)
        count := count + 1;
    END_WHILE;
END_PROGRAM;
```

Figure 17.1 – Errored-out ChatGPT-generated code

The red lines mean there are errors in the program. If you hover over the erroneous commands, you will see that the BEGIN keyword is not identified; in other words, it does not exist. This means that we need to do some massaging to make this work. The first thing we need to do is declare the count variable the system gave us, like the following:

```
PROGRAM PLC_PRG
VAR
    count : INT := 1; (* Counter variable *)
END_VAR
```

Once you have that variable implemented, you can move on to implementing the main logic, which is the following:

```
WHILE count <= 10 DO
    count := count + 1;
END_WHILE;
```

In this example, we gutted out the BEGIN and END_PROGRAM commands as well as the comments. In all, when the program is fully implemented, it should look like *Figure 17.2*:

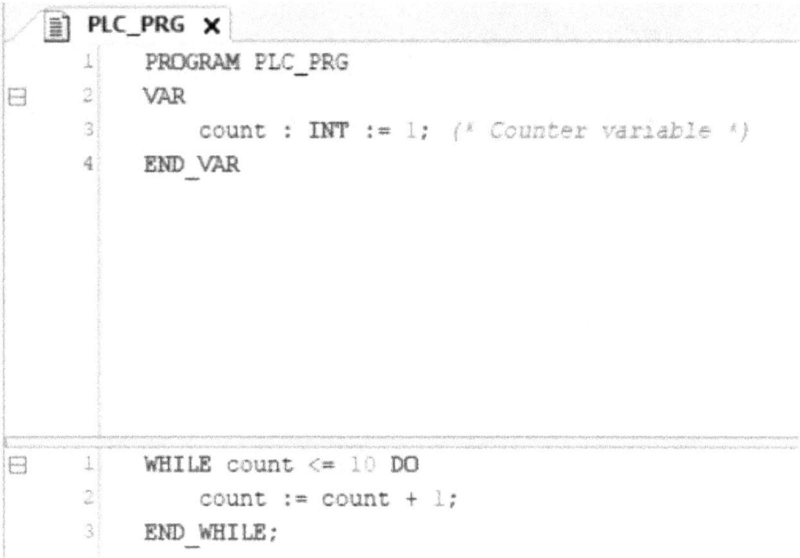

Figure 17.2 – Modified ChatGPT code

When this program is run, it should produce the output seen in *Figure 17.3*:

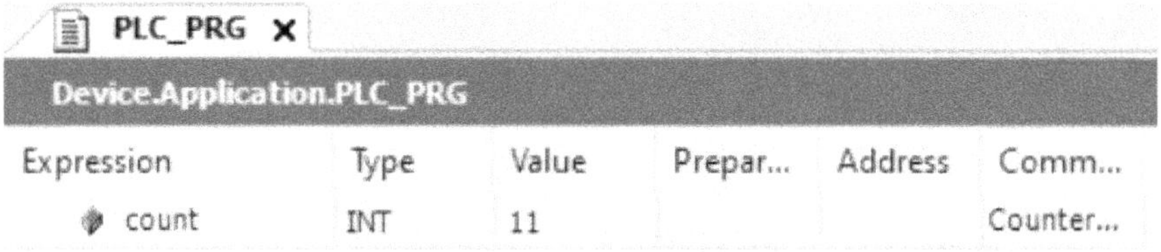

Figure 17.3 – Modified ChatGPT output code

Notice that the value counted to 11, not 10 as we wanted. This means that more modifications would be needed to make the program run as intended. We could fix this by having the system regenerate the code or fix it manually.

> **Note**
>
> This is where prompt tuning would normally come into play. A dedicated prompt engineer would rewrite the prompt until the system produced the correct code.

All in all, ChatGPT did produce code that was usable; we just had to modify it a little to make it work. If you think about it, we generated several lines of code, variables, and so on with a single line of text. Could we refine the prompt to produce 100% accurate code? Of course! Is doing this worth the added effort? Probably not.

This example shows that GenAI needs to be thought of as an assistant. Though the system didn't produce code that we could just plug in and run, it did produce code that could be easily tweaked to run. In all, we did save some time. With some knowledge of how to generate code with ChatGPT under our belt, we can move on to our final project!

Final project

For our final project, we're going to use ChatGPT to produce a program that will do the following:

- Set a `message` variable to `locked` when the temperature is above 100 degrees
- Set a `message` variable to `warning` when the temperature is between 95 and 100 degrees
- Set a `message` variable to `safe` when the temperature is below 95 degrees

Before you move on, try to write a prompt that can generate a program.

Prompt and code

For this project, we could use something like the following prompt:

```
Generate a Structured Text PLC program that will set a message to
locked when a temperature variable is above 100, set a message to
warning with the temperature variable is between 95 and 100, and set
the message to safe when the temperature is under 95.
```

The code that ChatGPT will produce is as follows:

```
PROGRAM TemperatureMonitoring
VAR
    Temperature : REAL := 0.0;      (* Temperature variable *)
    Message : STRING(20);           (* Message variable *)
END_VAR

PROGRAM TemperatureMonitoring
VAR
```

```
    Temperature : REAL := 0.0;(* Temperature variable *)
    Message : STRING(20);       (* Message variable *)
END_VAR
(* Main program *)
BEGIN
    (* Check temperature conditions *)
    IF Temperature > 100 THEN
        Message := 'Locked';            (* Set message to 'Locked' if
temperature is above 100 *)
    ELSIF Temperature >= 95 THEN
        Message := 'Warning';       (* Set message to 'Warning' if
temperature is between 95 and 100 *)
    ELSE
        Message := 'Safe';             (* Set message to 'Safe' if
temperature is under 95 *)
    END_IF;
END_PROGRAM;
```

Again, notice that the system produces code that will need to be massaged for it to work. To make this code compile and run, we're going to ignore everything but the IF block and the variables. In other words, the variable section should look like the following:

```
PROGRAM PLC_PRG
VAR
    Temperature : REAL := 0.0;
    Message : STRING(20);
END_VAR
```

The main body of the program should look like this:

```
IF Temperature > 100 THEN
    Message := 'Locked';
ELSIF Temperature >= 95 THEN
    Message := 'Warning';
ELSE
    Message := 'Safe';
END_IF;
```

Run the program and type in the value 101, and you should get the output seen in *Figure 17.4*:

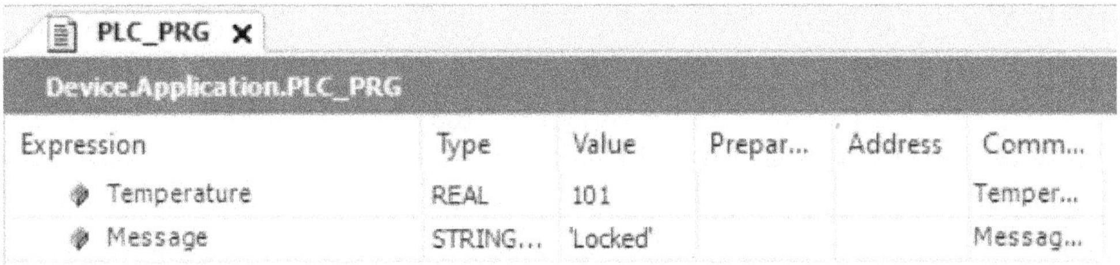

Figure 17.4 – Locked case

As can be seen, the message is set to locked, as expected. Next, we can try for a warning message by inputting 96 for the Temperature variable. *Figure 17.5* shows that the message is set correctly:

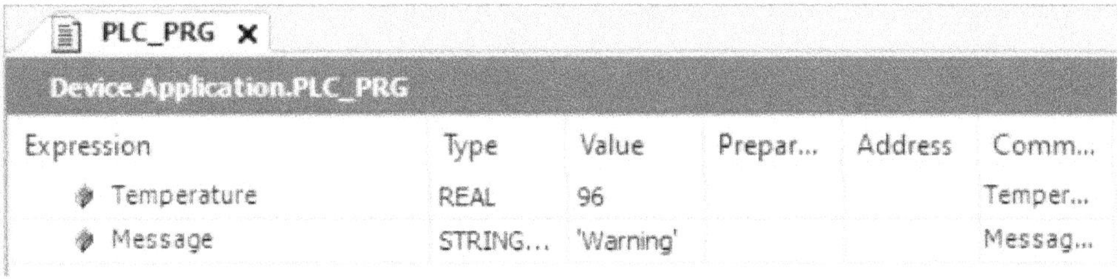

Figure 17.5 – Warning case

Finally, we can check the safe case by entering 76 into the Temperature variable:

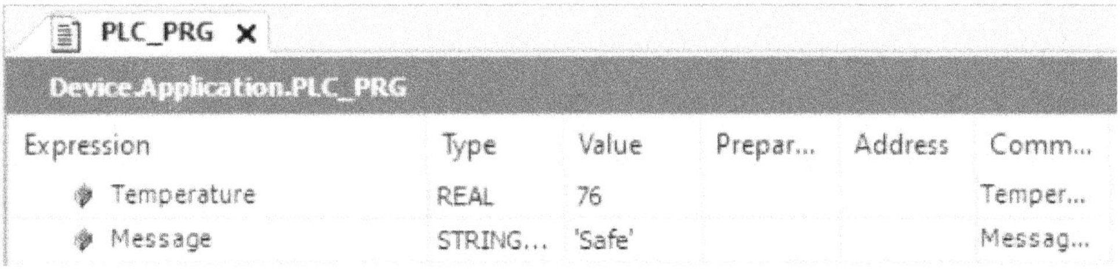

Figure 17.6 – Safe case

As can be seen, all the cases work. This means with minor omissions in the code, ChatGPT produced a working program.

Summary

In this chapter, we covered GenAI, ChatGPT, prompt engineering, and more. We also explored how well ChatGPT produces code. In all fairness, it produced code at the same level a junior PLC programmer would. It's not bad code, but it did take some massaging for it to work. As of now, and with the average prompt engineering skill level of most people, ChatGPT cannot be a stand-in for human programmers. In all fairness, a talented prompt writer with lots of time and an optimized system could probably produce grand code with the tool; however, for the average user, ChatGPT is not going to be a fill-in anytime soon. The key takeaway here is that **ChatGPT is a tool**. Just as a screwdriver is useless without a carpenter, ChatGPT and other systems are going to be equally useless without knowledgeable engineers driving. With the way the technology currently is, it is far better to think of ChatGPT and other AI systems as an assistant as opposed to an engineer.

By this point in the book, you should have a pretty good background in computer science, ST, and general PLC programming. As we conclude the book, the final chapter will be dedicated to putting everything together for our final project!

Questions

1. What is NLP?

2. What is GenAI?

3. What is a prompt?

4. What mindset should you have when developing a prompt?

5. What are the five main aspects of prompt engineering?

6. Will ChatGPT always produce 100% usable code?

7. What is the best way to think of GenAI?

8. State your opinion as to whether or not GenAI will replace human programmers.

Further reading

- *ChatGPT as an Assistant for PLC Programmers*:

  ```
  https://www.packtpub.com/article-hub/chatgpt-as-an-assistant-
  for-plc-programmers
  ```

- *ChatGPT for Ladder Logic*:

  ```
  https://www.packtpub.com/article-hub/chatgpt-for-ladder-logic
  ```

18

The Final Project – Programming a Simulated Robot

Congratulations – you have made it to the final chapter! This chapter is going to be the conclusion of our exploration into the basics of PLC and Structured Text programming. This chapter is going to be unique compared to the rest of the book in that it will be a comprehensive project that will draw on the main topics that were previously explored throughout this book. This means that a solid understanding of the rest of the book is going to be necessary to follow along. So, if you skipped to the end, you need to go back and explore the rest of the book before tackling this project.

This project is going to explore programming a simulated robot. In modern automation, more and more robots are being designed to interface directly with PLCs. This project is going to be a very simple programming scenario that one could face in the real world. Keep in mind that the code we build could be massaged enough to control an actual robot; this project is going to be a simulated example.

In automation, robots are pivotal pieces of equipment. One common application for robots is to sort parts. For example, many factories will integrate a robot into a machine that can sort parts, as we explored previously. The robot will move to a certain point, tag certain parts with a sticker, and send them down their respective lines to be rejected, reworked, or packaged. For our final project, we are going to create a simulated robot that can do this.

To accomplish this, we're going to break this chapter down into the following sections:

- Defining the scope of the project
- Defining the project's requirements
- Designing a flowchart to describe the overall process
- Designing pseudocode for the project

Finally, we're going to round out the chapter by implementing and testing the code for our simulated robot.

Technical requirements

To complete this chapter, you will need a working copy of CODESYS and a ChatGPT account. The code for this chapter will be available at the following URL:

```
https://github.com/PacktPublishing/PLCs-for-Beginners
```

This chapter will include most of what we explored in the previous chapters. This means it is advisable that you fully read and understand the material presented throughout the book. Of all the material this project will focus on, program design, flow control (mostly IF and CASE statements), and sorting. This chapter will also draw heavily on prompt engineering from the previous chapter.

Project scope

Recently, we have been tasked with creating an automation system that can integrate a robot into a sorting system. The system will be based around a robot; the system will sort parts if the largest part is 4 mm and the smallest is 3 mm, then the robot will tag the parts as acceptable and send them for packaging. If the largest part is between 4 mm and 4.5 mm and the smallest part is between 2 mm and 3 mm, the robot will tag them for rework and send them down the rework line. Any parts that do not meet either of these requirements will be tagged for recycling and sent down the recycle line.

The scope of this project is clear in what we need to do. We can easily build our requirements off of this scope. So, before you proceed, try to lay out the requirements on your own. When you're done, you can move on and compare your requirements to the actual requirements!

Project requirements

This system is going to have two parts that will work in unison: one is the robot and the other is the sorter. The sorter will need to sort the parts from least to greatest, and the robot will need to measure the smallest and largest parts. Once the parts are measured, it will tag the lots and send them down the proper line. This means we can set the following requirements:

- The sorter will need to sort all parts in an incoming lot from least to greatest

- Each lot will contain five parts

- The robot will need to measure the parts

    - If the largest part is 4 mm and the smallest is 3 mm, the robot will tag the part for packaging

    - If the largest part is between 4 mm and 4.5 mm and the smallest part is between 3 mm and 2 mm, then the robot will tag the parts for rework

    - If the largest part exceeds 4.5 mm and the smallest is under 2 mm, then the robot will tag the parts for recycling and send them for recycling

- If the part is tagged for packaging, send it down line 1

- If the part is tagged for rework, send it down line 2

- If the part is tagged for recycling, send it down line 3

- The system will need to reset itself and prepare for the next run

- Keep track of the number of rejected, packaged, and reworked orders

These are the high-level requirements. These requirements should be good enough for us to design a system to complete the task. Therefore, the next step is to flowchart the process.

Flowchart for the system

For this section, the first step in implementing a quality system is to first design a system flow based on the requirements. We can graphically represent this process in *Figure 18.1*. The process represented by the flowchart is straightforward and nearly mirrors the requirements. The process will start with sorting the parts from greatest to largest. After the parts are sorted, the robot will measure the largest and smallest parts. The system will then check the part range; if the parts are determined to be 4 mm and 3 mm for the largest and smallest part respectively, the robot will send them down the packaging line. If the parts are less than 4.5 but larger than 4 mm for the largest and are less than 3 mm but more than 2 mm for the smallest, the part will be tagged as needing rework and sent down that line:

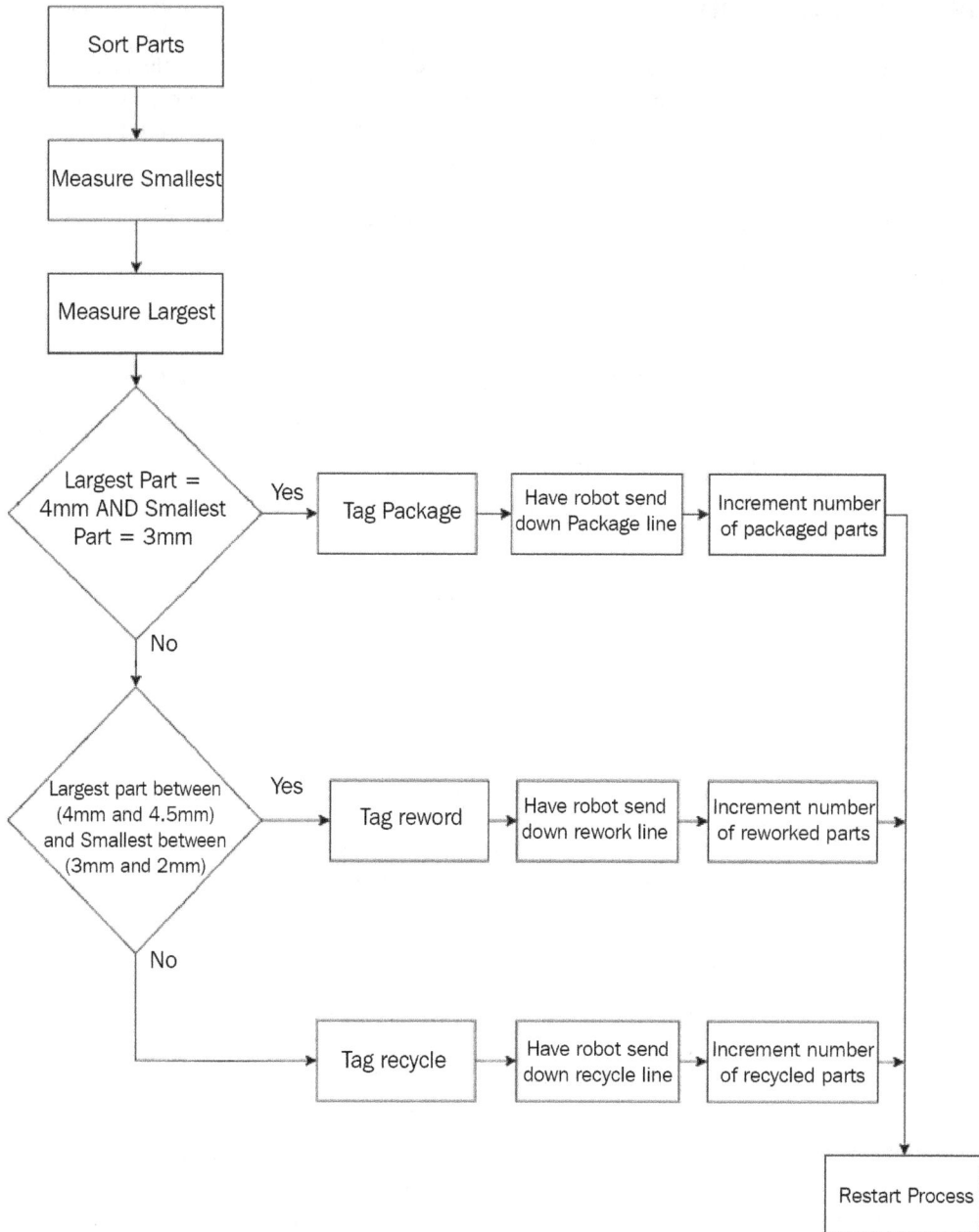

Figure 18.1 – System process

From the preceding diagram, this system is a great candidate for a state machine. The next step in designing the system is to work out the pseudocode.

Designing the pseudocode

The following pseudocode can be used as a basis for the code:

```
Counter(increment)
If start = True Then
CASE
1: //sort parts
        Use bubble sort to sort parts
        Robot position = "home pos"
        set case 2
2: //measure parts
        Direct robot to measure parts
        Robot position = "measure rack"
        set case 3
3: //tag parts
    Robot position = "tag position"
        If largePart = 4mm AND smallPart = 3mm Then
            Tag part for package
        Else if largePart (between 4.5mm and 4mm) OR
                    SmallPart(between 3mm and 2mm) Then
            Tag part for package
            Else
            Tag for recycle
        Set case 4
4: //send down line
            If tag = package Then
                Send to package line
                Increment packaged number
            Else if tag = rework Then
                Send to rework line
                Increment rework number
            Else if tag = recycle Then
                Send to recycle line
                Increment recycle number
        Robot position = "move package"
    Set case 5:
5: //restart
        Start = false
        Robot position = "home"
        Set case 1
```

This pseudocode will provide a quality roadmap for our working program. To prevent the program from running continuously, we embedded the state machine in an `IF` block that is controlled by a `start` variable. This variable will simulate a button that an operator will press to start the process. To keep track of the number of processes run, we're going to use a `CTU` instruction to keep track of that.

The only code that could be tricky to implement is the bubble sort algorithm that we explored in *Chapter 14*. However, to make things easier, we're going to have ChatGPT implement that algorithm for us. Now that we have a design for our process and our program, we can move on to implementing the code!

Implementing the program

The first step on the road to turning our designs into a working program is to start with getting the code for the sorting algorithm ready. As stated before, we're going to use ChatGPT to quickly generate code for the sorting algorithms.

Using ChatGPT to implement bubble sort

The pseudocode stated that the sorting algorithm we are going to use is the bubble sort algorithm. As was explored in *Chapter 14*, the bubble sort algorithm isn't necessarily hard to implement, but it can be tedious. So, to speed things along, we're going to use ChatGPT to generate code for us.

Before using the following prompt, try to design the prompt on your own to get a bubble sort algorithm. To generate our bubble sort algorithm, we're going to use the following prompt:

```
Generate a bubble sort algorithm in Structured Text for an array with
five elements.
```

This prompt will produce the following code:

```
PROGRAM BubbleSort
VAR
    i, j, temp : INT;
    arrayToSort : ARRAY[1..5] OF INT := [5, 3, 1, 4, 2];   (* Example
array to be sorted *)
END_VAR

(* Bubble sort algorithm *)
FOR i := 1 TO 4 DO
    FOR j := 1 TO (5 - i) DO
        IF arrayToSort[j] > arrayToSort[j + 1] THEN
            (* Swap elements *)
            temp := arrayToSort[j];
            arrayToSort[j] := arrayToSort[j + 1];
            arrayToSort[j + 1] := temp;
```

```
        END_IF;
    END_FOR;
END_FOR;
```

The only things we need to change are the data types of the `array` variable and the `temp` variable. Since the requirements state that there is one condition that requires decimal values, we need to change those values to `REAL`. The only thing that will need to be modified is the values in the array. To use this code, all we must do is use the variables and pass that labeled bubble sort algorithm into the first case in the `CASE` statement. With the bubble sort portion of the program squared away, we can move on to the code implementation, starting with the variables.

Challenge

Notice that ChatGPT generated a code block that sorted integers. For our purposes, this is fine because we can easily modify the code to fit our needs. However, ChatGPT can also easily generate the correct code with minor tweaks to the prompt. As a challenge, rewrite the prompt to accommodate the decimal values that will be needed.

Final project variables

This program will require the following variables:

```
PROGRAM PLC_PRG
VAR
    //start process
    start        : BOOL := FALSE;

    //robot pos
    robotPos     : WSTRING;

    //state of the process
    processState : INT;

    //run counter
    partRun      : CTU;
    partRunCount : INT;

    //smallest
    smallestPart : REAL;
    largestPart  : REAL;

    //tag
    tag : WSTRING;

    //line
```

```
line      : WSTRING;

//parts stats
numOfRecycledParts : INT := 0;
numOfReworkedParts : INT := 0;
numOfPackagedParts : INT := 0;

//bubble sort vars
i, j : INT;
temp : REAL;
arrayToSort : ARRAY[1..5] OF REAL := [3, 3, 3, 4, 3];

END_VAR
```

The variables' functions are noted in the source code. This variable list also includes the variables that were generated with ChatGPT with their modified data type. After you implement the variables, you can move on to implementing the logic.

Main program logic

The main program logic should look like the following:

```
partRun(CU := start);
partRunCount := partRun.CV;
IF start = TRUE THEN
    CASE processState OF
    1: //sort
        robotPos := "home pos";
        FOR i := 1 TO 4 DO
            FOR j := 1 TO (5 - i) DO
                IF arrayToSort[j] > arrayToSort[j + 1] THEN
                    (* Swap elements *)
                    temp := arrayToSort[j];
                    arrayToSort[j] := arrayToSort[j + 1];
                    arrayToSort[j + 1] := temp;
                END_IF;
            END_FOR
        END_FOR
        processState := 2;

    2:// measure parts
        robotPos := "measure rack";
        smallestPart := arrayToSort[1];
        largestPart  := arraytoSort[5];
```

```
            processState := 3;

    3:// tag parts
        robotPos := "tag position";
        IF largestPart = 4 AND smallestPart = 3 THEN
            tag := "Package";
        ELSIF (largestPart > 4 AND largestPart < 4.5) OR
(smallestPart < 3 AND smallestPart > 2) THEN
            tag := "Rework";
        ELSE
            tag := "Recycle";
        END_IF

        processState := 4;

    4: //send down line
        IF tag = "Package" THEN
            line := "Package Line";
            numOfPackagedParts := numOfPackagedParts + 1;
        ELSIF tag = "Rework" THEN
            line := "Rework Line";
            numOfReworkedParts := numOfReworkedParts + 1;
        ELSIF tag = "Recycle" THEN
            line := "Recycle Line";
            numOfRecycledParts := numOfRecycledParts + 1;
        END_IF
        robotPos := "move package";
        processState := 5;
    5: //restart
        start := FALSE;
        robotPos := "home";
        processState := 1;
    END_CASE
 END_IF
```

This code closely mirrors the pseudocode. As stated before, this overarching architecture utilizes a state machine, which is represented in the CASE statement. The whole system is dependent on the start variable, which simulates a button press by the operator. When this variable is toggled to True, the number of runs represented by the partRunCount variable will increase by one. As each subprocess, such as the tag or measure process, finishes, the processState variable will also change to the next phase in the process. With the code and logic of how the program was established, we can move on to testing it.

Testing the program

For the very first test run, we're going to test the package functionality. For this, we can use the array we set in the variable section of the code.

Testing the packaging function

For this test, simply set start to True and processState to 1:

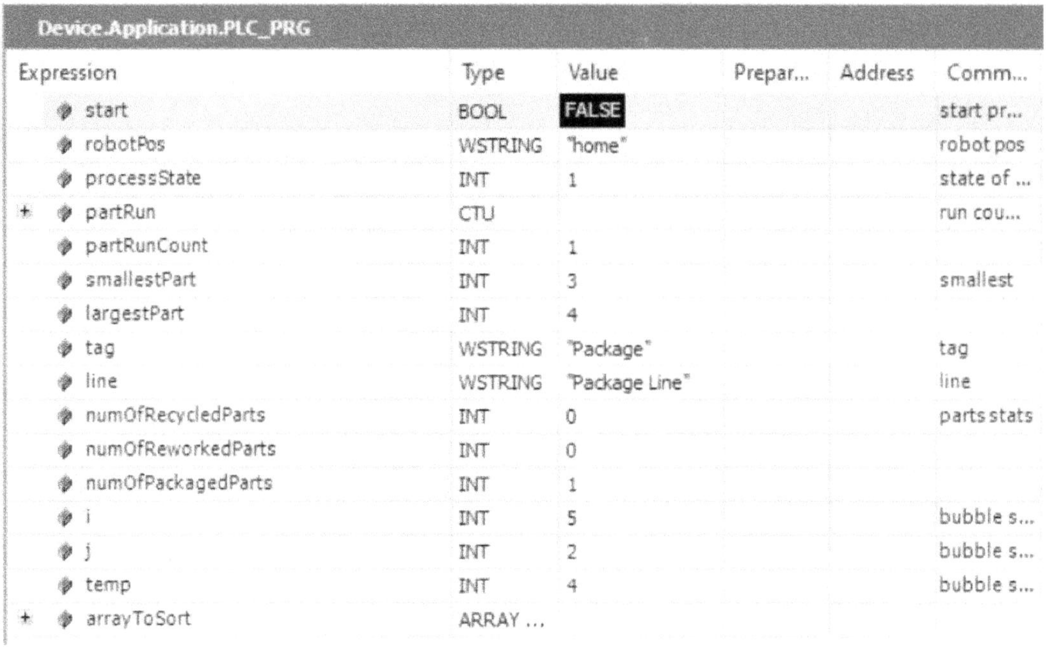

Expression	Type	Value	Prepar...	Address	Comm...
Device.Application.PLC_PRG					
start	BOOL	FALSE			start pr...
robotPos	WSTRING	"home"			robot pos
processState	INT	1			state of ...
partRun	CTU				run cou...
partRunCount	INT	1			
smallestPart	INT	3			smallest
largestPart	INT	4			
tag	WSTRING	"Package"			tag
line	WSTRING	"Package Line"			line
numOfRecycledParts	INT	0			parts stats
numOfReworkedParts	INT	0			
numOfPackagedParts	INT	1			
i	INT	5			bubble s...
j	INT	2			bubble s...
temp	INT	4			bubble s...
arrayToSort	ARRAY ...				

Figure 18.2 – Packaging output

As can be seen in *Figure 18.2*, the tag and line are both correct. That is, they are both the package line and tag. Another important attribute is partRunCount. This value started off at 0 and was increased to 1 as expected. Finally, we can move on to the robotPos variable, which says home. This means that a theoretical robot resets itself and is not waiting for its next run. In all, the packaging functionality is working as expected, so we can now move on to test the rework functionality of the machine!

Testing the rework function

To test this feature, we're going to use the following array:

```
arrayToSort : ARRAY[1..5] OF REAL := [2, 2, 3, 4.4, 3];
```

When you run the program, ensure you log in with download, and you should be met with the following:

Device.Application.PLC_PRG					
Expression	Type	Value	Prepar...	Address	Comm...
🔷 start	BOOL	FALSE			start pr...
🔷 robotPos	WSTRING	"home"			robot pos
🔷 processState	INT	1			state of ...
+ 🔷 partRun	CTU				run cou...
🔷 partRunCount	INT	1			
🔷 smallestPart	REAL	2			smallest
🔷 largestPart	REAL	4.4			
🔷 tag	WSTRING	"Rework"			tag
🔷 line	WSTRING	"Rework Line"			line
🔷 numOfRecycledParts	INT	0			parts stats
🔷 numOfReworkedParts	INT	1			
🔷 numOfPackagedParts	INT	0			
🔷 i	INT	5			bubble s...
🔷 j	INT	2			bubble s...
🔷 temp	REAL	4.4			
+ 🔷 arrayToSort	ARRAY ...				

Figure 18.3 – Rework functionality

As can be seen in *Figure 18.3*, our part sizes triggered the robot to send the part down the rework line. This is the expected behavior, which means we can move on to testing the final functionality: the recycle functionality!

Testing the recycling function

The first step in testing the recycling functionality is to adjust the array to trigger the recycling process. To do this, we can adjust the array to the following:

```
arrayToSort : ARRAY[1..5] OF INT := [1, 1, 3, 4, 5];
```

As can be seen, in this array, the smallest number is 1 and the largest is 5. This will trigger multiple cases in the IF statements that will trigger the recycling functionality. When the program is run, ensure you log in with download, and you should be met with the output shown in *Figure 18.4*:

Expression	Type	Value	Prepar...	Address	Comm...
Device.Application.PLC_PRG					
🔷 start	BOOL	FALSE			start pr...
🔷 robotPos	WSTRING	"home"			robot pos
🔷 processState	INT	1			state of ...
⊞ 🔷 partRun	CTU				run cou...
🔷 partRunCount	INT	1			
🔷 smallestPart	INT	1			smallest
🔷 largestPart	INT	5			
🔷 tag	WSTRING	"Recycle"			tag
🔷 line	WSTRING	"Recycle Line"			line
🔷 numOfRecycledParts	INT	1			parts stats
🔷 numOfReworkedParts	INT	0			
🔷 numOfPackagedParts	INT	0			
🔷 i	INT	5			bubble s...
🔷 j	INT	2			bubble s...
🔷 temp	INT	0			bubble s...
⊞ 🔷 arrayToSort	ARRAY ...				

Figure 18.4 – Recycling line

Figure 18.4 shows that the program is triggering the recycling process as expected and the robot has reset itself for a new run!

Challenge

In terms of the simulated robot, there is a robot Pos variable that will track where the robot is in the process. This variable will change rapidly, and the changes may not be noticed during runtime. For this challenge, add in some pauses using a TOF instruction to add a pause in between process changes.

Summary

Congratulations – you have now finished the book! By this point in the book, you should have a good grasp of ST, which is something that typical PLC programmers lack. You should also have a quality grasp of computer science topics in general. This book has also introduced you to new and rarely considered concepts in PLC programming, such as **generative AI** (**GenAI**) and security respectively.

The main takeaway from this book is that PLC programming is way more than just basic Ladder Logic. Though Ladder Logic can get you through the day in some jobs, your flexibility to meet new and ever-increasing challenges that PLC programmers are either facing or will face soon will require you to understand not only Structured Text but also how to write safe, efficient, and malleable code. In all, this book has been an introduction to those concepts. However, modern PLC programming has much more to offer, such as the very rich world of **object-oriented programming** (**OOP**). Therefore, it is highly recommended that you continue your learning with the second part of this book, *Mastering PLC Programming – The Software Engineering Survival Guide to Automation Programming*, which will take the concepts that you learned here, greatly expand on them, and introduce you to what is the next generation of automation programming.

Assessments

Chapter 1

1. What are three use cases for a PLC?

    There are many uses cases; however, common uses cases are as follows:

    - Streetlights
    - Amusement parks
    - Factories
    - Cranes
    - Nuclear reactors
    - Space launch systems
    - Dams

2. Can a PLC be used in a space launch system?

    Yes.

3. Why is computer science important to an automation programmer?

    Computer science is the study of writing software for digital systems such as PLCs and other devices. Therefore, it can improve the quality of your PLC code!

4. What are two use cases for a microcontroller?

    - Toys
    - Consumer electronics

5. Name three emerging technologies for industrial automation.

    - Cloud computing
    - IoT
    - AI/Machine Learning

6. What is computer science?

    The study of computation such as information processing.

7. Why should automation programmers care about computer science?

    Computer science can help PLC programmers write safer, faster, and more effective software.

8. Name three common microcontrollers.

    - PICs

    - AVRs

    - Arduinos

9. What are some common microcontroller programming languages?

    There are many like:

    - The Arduino programming language

    - A BASIC dialect

    - A C dialect

Chapter 2

1. Name three types of PLC modules.

    There are many types of modules such as:

    - A CPU module

    - A power supply

    - A analog input

    - A digital input

    - A chassis

    - A safety variant of any of these

2. What is the difference between a regular PLC module and a safety module?

    A safety module is designed to fail gracefully.

3. What is a stepper motor?

    A motor whose angle is controlled by pulses.

4. What is a servo motor?

    A closed loop motor.

5. How many pulses are required to move a stepper 180 degrees if the resolution is 1.8?

   100.

6. What is a motor encoder?

   A device that gives positional feedback.

7. What is a motor drive?

   A device that sends control signals to the motor.

8. What is the difference between an analog module and a digital module?

   An analog module will take in or send out a range of values. A digital module will only produce a full on or fully off condition.

9. What is a discrete module?

   A module whose input or output is either fully on or fully off.

10. What is a BOM?

   Bill Of Material.

11. What type of module should an E-Stop be wired into?

   Safety module.

12. What is an analog input?

   An I/O module that takes a range of signal inputs usually from temperature, pressure, weight, and various other sensors.

13. What is a discrete input?

   A digital input.

Chapter 3

1. What is the difference between a compiler and an interpreter?

   A compiler will translate the code base all at one time and find errors before the program can run. An interpreter will translate a program line-by-line.

2. What is IEC 61131-3?

   A standard that governs the programming languages for compliant PLCs.

3. What is a machine instruction?

   Typically defined as an instruction that a family of CPUs use to carry out tasks.

4. What is a programming paradigm?

   A way in which code is architected and implemented. Examples are object-oriented, procedural, and so on.

5. Does IEC 61131-3 support OOP?

   Yes.

6. What is OOP?

   Object-Oriented Programming.

7. Is IEC 61131-3-compliant code always portable between devices?

   No.

8. Write an algorithm for withdrawing $20 from an ATM.

   A. Enter credit card

   B. Enter pin number

   C. Press withdraw button

   D. Input $20

   E. Take money from ATM

   F. Put money in wallet

   G. Walk off

9. What is a language translator?

   A program that translates human-readable code into machine readable instructions.

10. What are the two types of language translators explored so far?

    - Compiler

    - Interpreter

11. What are the languages that the IEC 61131-3 supports?

    - Structured Text

    - Sequential Function Charts

    - Ladder Logic

    - Instruction List

    - Function Block Diagram

12. In what direction does a program flow?

Top to bottom.

13. What is syntax?

The grammar of a programming language

14. What are keywords?

Reserved words that trigger actions when coupled with the correct syntax.

Chapter 4

1. How does an SSD work?

SSDs use flash memory to store data.

2. What is a memory address?

A computer readable name for a location in memory.

3. What is an example of an obsolete storage device?

A floppy drive.

4. Name two modern storage devices.

- An SSD drive

- A USB drive

5. What is cloud storage?

Computer storage the is usually offered by a third-party that utilizes the internet as a transfer medium.

6. What are two drawbacks to cloud storage?

- Requires a connection to the storage medium (typically an internet connection)

- Can be costly

7. What is a memory block?

A unit of computer memory

8. What is volatile memory?

Memory that will lose data when power is withdrawn from it.

9. What is non-volatile memory?

Memory that will not lose data when power is withdrawn from it.

10. What kind of memory is ROM?

    Non-volatile memory.

11. What kind of memory is RAM?

    Volatile memory.

12. What is storage?

    Long term retention of data.

13. What does RAM stand for?

    Read Access Memory

14. What does ROM stand for?

    Read Only Memory.

Chapter 5

1. What is a flowchart?

    A graphical representation of a program's flow.

2. What is pseudocode?

    A language-based representation of a program using common language semantics.

3. What are the main differences between pseudocode and flowcharts?

    * Pseudocode is language based

    * Flowcharts are graphical

4. List three programs that can be used to write pseudocode.

    A programmer could use:

    * Microsoft Word

    * Notepad

    * Notepad++

5. List three programs that can be used to draw flowcharts.

    * Visio

    * Draw.io

    * Flowgorithm

6. Why use a flowchart?

    Its graphical nature can help:

    - Visualize the flow of the program

    - Help find redundancies in the program

    - Help find pinpoints in the program

7. Can you draw flowcharts by hand?

    Yes.

8. What is a common interview technique that requires candidates to draw a design on a whiteboard?

    Whiteboarding.

9. What are the two functions of the diamond symbol in a flowchart?

    - Though it will vary from program to program a diamond symbol can represent an IF statement or a loop.

    - For the remainder of this book a simple square will be used to alleviate confusion with IF statements.

10. Design a robot stop system that sequentially turns off all the systems we turned on in the final project.

    There is no right or wrong answer as this will vary from programmer to programmer.

Chapter 6

1. Solve the following equation: $(A + B) + A$ where $A = 1$ and $B = 0$.

    1.

2. Write the truth table for question 1.

A	B	(A+B) + A
1	1	1
1	0	1
0	1	1
0	0	0

3. What is the truth table for the AND operator?

A	B	Output
1	1	1
1	0	0
0	1	0
0	0	0

4. What does 0 represent?

    Off or false.

5. What does 1 represent?

    On or true

Chapter 7

1. What are two programming languages that are similar to ST?

    In terms of PLC languages, the closes one would be Instruction List. Outside of PLC languages, one could say C/C++, Java, C#, Visual Basic, Ada, and so on.

2. What are the two sections of the PLC_PRG file?

    -   The variable section
    -   The logic sections

3. Why would you use ST over LL?

    ST provides and easy to use read interface that makes following the flow of a program much easier and can require less overall code.

4. What does the Value column represent?

    It shows the current value a variable is set to.

5. What is the process to run a PLC program in CODESYS?

    A. Put CODESYS in simulation mode.
    B. Click the login button.
    C. Press the play button.

Chapter 8

1. What data type is used to store whole numbers?

    INT

2. What is a floating-point number?

    A decimal number.

3. What data type is used to store a floating-point number?

    REAL or LREAL

4. What data type would you use to store the result of 3/2?

    REAL or LREAL

5. What is a strongly typed language?

    A language that evaluates data types before an operation is preformed.

6. What is a data type?

    The type of data a variable can hold. For example, a whole number or string of characters.

7. What is a weakly typed language?

    A language that does not evaluate data types before operations are preformed .

8. What is a dynamically typed language?

    A language that figures out the data type for you.

9. What is a statically typed language?

    A language that requires you to explicitly declare the data type of a variable.

10. Is CODESYS dynamically or statically typed?

    Statically typed.

Chapter 9

1. What does the floor function do?

    Return the next lowest whole number.

2. What is the ABS of -3?

    3

3. What is the ATAN function?

    Inverse of the TAN function.

4. How would you write a program that can calculate a quadratic equation?

```
PROGRAM PLC_PRG
VAR
    a : INT;
    b : INT;
    c : INT;
    root1 : REAL;
    root2 : REAL;
END_VAR
root1 := ((-1 * b) + SQRT(EXPT(b,2) - (4*a*c))) / (2*a);
root2 := ((-1 * b) - SQRT(EXPT(b,2) - (4*a*c))) / (2*a);
```

5. How would you write a program that can calculate 4 to the power of 3?

```
EXPT(4,3)
```

6. What is the order of operations for a program?

    Parentheses, exponents, multiplication, division, addition, and subtraction.

7. What is the assignment operator?

```
:=
```

Chapter 10

1. What is a function block?

    A digital blueprint that is equivalent to a class in a general-purpose programming language.

2. Can a function live in a function block?

    Technically, no. Methods which are equivalent to functions live in function blocks.

3. Do all PLC programming systems support function blocks?

    Generally, yes.

4. Name three common function blocks in CODESYS.

    - TON

    - TOF

    - CTU

5. Do all function blocks have the same inputs and outputs?

No.

6. What is the main difference between a TON and a TOF timer?

TON on hold a bit off for a given amount of time while a TOF will be on for a given amount of time.

7. What is the main difference between a timer and a counter?

A counter will increment with a rising edge while a timer will toggle a bit after a give period of time.

Chapter 11

1. What direction does a program always flow in?

Top to bottom.

2. When will an IF statement run?

When the logical expression it contains evaluates to true.

3. Will an IF statement be executed if the condition evaluates to false?

No.

4. What is a state machine?

A programming structure where an input dictates a machine state.

5. What is a common way to implement state machines?

A CASE statement.

6. What is a CASE statement?

A CASE statement is an easy way to implement a series of branches in a program.

7. What is the CASE statement syntax?

```
CASE input OF
    0:
        //code
    1:
        //code
    2:
        //code
END_CASE
```

8. Can you have an IF statement inside of a CASE statement?

    Yes.

9. What is the major difference between an IF and a CASE statement?

    An IF statement can only evaluate one condition while a CASE can evaluate many.

10. How can flowcharting help with flow control?

    Allows you to graphically depict the flow of a program and its many branches.

11. What is the difference between "less than or equal to" or just "less than" instructions?

    Less than or equal to will evaluate to true when the value on the left is less than or equal to the number on the right of the symbol. Less than will only evaluate to true when the number on the left is less than the number on the right of the symbol.

12. What symbol is used to test for "not equals" in an IF statement?

    ```
    < >
    ```

13. What symbol is used to test for "greater than" in an IF statement?

    ```
    >=
    ```

14. What is the minimum amount of code needed for an IF statement?

    ```
    IF <expression> THEN
        //code
    END_IF
    ```

15. What happens to the code in an IF statement when the expression evaluates to true?

    The code embedded in the IF statement will run.

Chapter 12

1. What is the difference between AND and OR operators?

    For the expression that contains the AND operator to evaluate to true all conditions must evaluate to true. For the expression that contains the OR operator to evaluate to true only one condition must evaluate to true.

2. What is the NOT operator used for?

    Use to invert the logical expression's output.

3. What is a nested IF statement?

    An IF statement inside of another IF statement or CASE statement.

4. What's the difference between ELSE and ELSIF statements?

    The ELSEIF is like an extra IF statement. The ELSE is a general-purpose conditional branch that will execute when all other IF and ELSEIF statements evaluate to false.

5. Can a CASE statement use an ELSE statement?

    Yes.

6. Can an ELSE statement accept a logical expression?

    No.

7. What is the truth table for an XOR statement?

A	B	Output
1	1	0
0	1	1
1	0	1
0	0	0

Chapter 13

1. What type of loop is a WHILE loop?

    Precheck loop.

2. What type of loop is a FOR loop?

    Counter loop.

3. What type of loop is a REPEAT loop?

    Post check loop.

4. What is a post check loop?

    A loop that is guaranteed at least one iteration. Implemented with REPEAT.

5. What is a precheck loop?

    A loop that is not guaranteed an iteration. Implemented with WHILE.

6. What is a counter loop?

    A loop that will iterate a given number of times. Implemented with FOR.

7. How many iterations are you guaranteed with a WHILE loop?

None.

8. How many iterations are you guaranteed with a REPEAT loop?

1.

9. When will a FOR loop terminate?

When a certain number of iterations have been completed.

10. What does the EXIT command do?

Prematurely terminates a loop.

11. Does the FOR loop use a custom logic statement to terminate?

No.

Chapter 14

1. What is an algorithm?

A predefined set of computer instructions that complete a specific task.

2. How is pseudocode used with algorithms?

Pseudocode is often used to present an algorithm in a language agnostic manner.

3. What does $O(1)$ mean?

Constant time complexity.

4. What is a more efficient time complexity: $O(n*log(n))$ or $O(n^2)$?

$O(n*log(n))$

5. How does merge sort work?

Merge sort uses a divide and conquer methodology.

6. Which is more efficient: bubble sort or merge sort?

Merge sort.

7. For an array (0..238), how do you retrieve the first element in the array?

Access element 0.

8. For an array (1..299), how do you retrieve the last element in the array?

Access element 299.

9. How do you calculate the number of elements in an array?

   `SIZEOF(Array)/SIZEOF(Array[1])`

10. Name three sorting algorithms.

    - Bubble sort

    - Merge sort

    - Insertion sort

11. Name three areas where an algorithm can be used.

    - AI/Machine Learning

    - Security

    - Sorting

    - Encryption/decryption

Chapter 15

1. Is the activation code in the final project a weak password?

   Yes.

2. What would be a better activation code for the final project?

   Anything that would have 8 characters or more and contain letters, numbers, and special characters. For example, `A1@Y7^2!`.

3. What is social engineering?

   Usually, a link or electronic message meant to dupe an individual into providing sensitive information.

4. What is a brute force attack?

   A password cracking attack that uses a program to try different combination of characters to guess a password.

5. What is a dictionary attack?

   A password cracking attack that uses a program that reads from a file to try out different passwords.

6. What is an air-gapped system?

   A network or computer that is not connected to outside networks like the internet.

7. Should you air-gap a system?

   When possible.

8. What are the triple As of security?

   - Authentication

   - Authorization

   - Accounting

9. What is a threat?

   A potential danger in a program.

10. What is a vulnerability?

    A weak spot in a program's security.

11. If you find a USB drive, should you plug it into your network? Why/why not?

    No. The USB drive may contain malware that can load itself onto the system.

12. Can a PLC be infected with malware?

    Yes, but depending on the brand it may require a specialized piece of malware.

13. Should you secure a PLC or the network the PLC is on?

    Yes.

14. Should you store passwords in the PLC or HMI, considering neither is attached to a network?

    Varies. Typically, the HMI will be harder to access; however, you can also store the password in a PLC.

15. What is an insider threat?

    A person that is authorized to use a system or access data and carries out nefarious actions knowingly or not.

16. What is a script kiddy?

    Unskilled individuals that attempt to gain unauthorized access to a network or system.

17. Who can be a hacker?

    Technically, anyone that is a computer professional or enthusiast.

18. What is the difference between a hacker and a cracker?

    A hacker is a computer enthusiast or professional while a cracker is a person that attempts to gain unauthorized access to a network or system.

19. What is a hacktivist?

    A group of individuals that break into networks or systems for a cause.

20. What is a nation-state attacker?

    Usually, a government or government agency with advanced resources.

Chapter 16

1. What is a symptom of a bad power supply?

   - Random reboots
   - Random shutdowns
   - Fluctuating power output
   - Machine not turning on

2. What are a few tools you should have in your toolkit?

   - Screwdrivers
   - Pliers
   - Thermal gun
   - Cables
   - Wire strippers
   - Multimeter
   - Wire crimps/strippers
   - Flashlights
   - Computer with necessary programming software installed on it.

3. What does the ping command do?

   It sees if the target device is responsive over the network.

4. What is a symptom of a network issue?

   Can't communicate with other devices on the network.

5. What is a possible symptom of a failed CMOS battery?

- Warning lights

- Blinking lights

- Wrong date/time

- Beeping sounds

- PLC shutdown

- Error messages

6. What is version control?

Special software that allows you to keep different versions of a program.

7. Name two benefits of version control.

Can be any of the following:

- Promote collaboration

- Keep code in a centralized connection

- Prevent the loss of older iterations of the code

- Prevent the loss of the current version of the code

8. What is print debugging?

A debugging technique where you program messages throughout the code base to see how far the program can execute and see its current location.

9. What is the difference between forcing and writing a value?

A variable this is written can be easily written over; thereby, changing its value. A variable that is forced must be forced to change its value, making it a more complex process.

10. How should you respond to broken software?

Gather information about the situation. Typically, the issue will stem from another cause such as faulty hardware or an update.

Chapter 17

1. What is NLP?

   Natural Language Processing.

2. What is GenAI?

   AI systems that can generate content.

3. What is a prompt?

   A query to the AI system that asks it to perform a task or answer a question.

4. What mindset should you have when developing a prompt?

   You should have a question framework in mind when developing a prompt.

5. What are the five main aspects of prompt engineering?

   For this question write a prompt and ask ChatGPT!

6. Will ChatGPT always produce 100% usable code?

   Sometimes. Most of the time the code will need to be massaged.

7. What is the best way to think of GenAI?

   As an assistant.

8. State your opinion as to whether or not GenAI will replace human programmers.

   State *YOUR* opinion!

Index

H

I

K

‹packt›

packtpub.com

Subscribe to our online digital library for full access to over 7,000 books and videos, as well as industry leading tools to help you plan your personal development and advance your career. For more information, please visit our website.

Why subscribe?

- Spend less time learning and more time coding with practical eBooks and Videos from over 4,000 industry professionals

- Improve your learning with Skill Plans built especially for you

- Get a free eBook or video every month

- Fully searchable for easy access to vital information

- Copy and paste, print, and bookmark content

Did you know that Packt offers eBook versions of every book published, with PDF and ePub files available? You can upgrade to the eBook version at packtpub.com and as a print book customer, you are entitled to a discount on the eBook copy. Get in touch with us at customercare@packtpub.com for more details.

At www.packtpub.com, you can also read a collection of free technical articles, sign up for a range of free newsletters, and receive exclusive discounts and offers on Packt books and eBooks.

Other Books You May Enjoy

If you enjoyed this book, you may be interested in these other books by Packt:

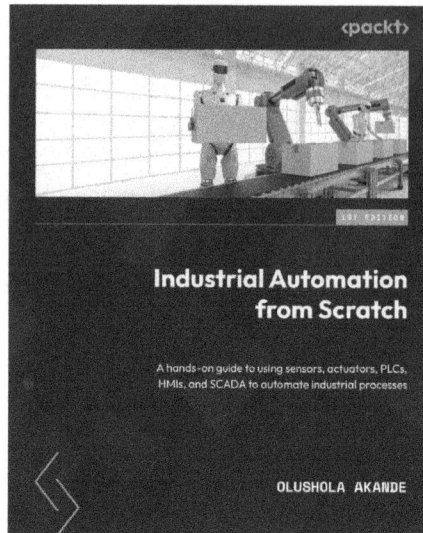

Industrial Automation from Scratch

Olushola Akande

ISBN: 978-1-80056-938-6

- Get to grips with the essentials of industrial automation and control
- Find out how to use industry-based sensors and actuators
- Know about the AC, DC, servo, and stepper motors
- Get a solid understanding of VFDs, PLCs, HMIs, and SCADA and their applications
- Explore hands-on process control systems including analog signal processing with PLCs
- Get familiarized with industrial network and communication protocols, wired and wireless networks, and 5G
- Explore current trends in manufacturing such as smart factory, IoT, AI, and robotics

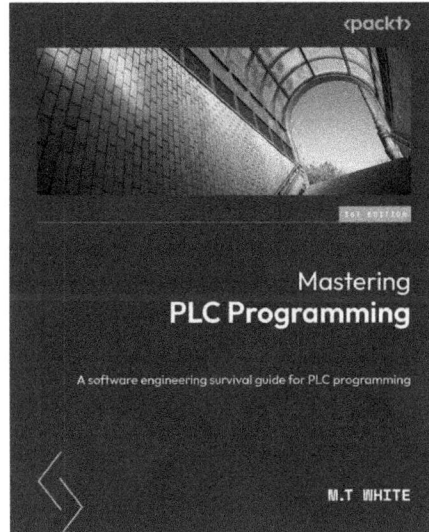

Mastering PLC Programming

Mason White

ISBN: 978-1-80461-288-0

- Find out how to write PLC programs using advanced programming techniques
- Explore OOP concepts for PLC programming
- Delve into software engineering topics such as libraries and SOLID programming
- Explore HMIs, HMI controls, HMI layouts, and alarms
- Create an HMI project and attach it to a PLC in CODESYS
- Gain hands-on experience by building simulated PLC and HMI projects

Packt is searching for authors like you

If you're interested in becoming an author for Packt, please visit `authors.packtpub.com` and apply today. We have worked with thousands of developers and tech professionals, just like you, to help them share their insight with the global tech community. You can make a general application, apply for a specific hot topic that we are recruiting an author for, or submit your own idea.

Share your thoughts

Now you've finished *PLCs for Beginners*, we'd love to hear your thoughts! Scan the QR code below to go straight to the Amazon review page for this book and share your feedback or leave a review on the site that you purchased it from.

`https://packt.link/r/1803230932`

Your review is important to us and the tech community and will help us make sure we're delivering excellent quality content.

Download a free PDF copy of this book

Thanks for purchasing this book!

Do you like to read on the go but are unable to carry your print books everywhere?

Is your eBook purchase not compatible with the device of your choice?

Don't worry, now with every Packt book you get a DRM-free PDF version of that book at no cost.

Read anywhere, any place, on any device. Search, copy, and paste code from your favorite technical books directly into your application.

The perks don't stop there, you can get exclusive access to discounts, newsletters, and great free content in your inbox daily

Follow these simple steps to get the benefits:

1. Scan the QR code or visit the link below

https://packt.link/free-ebook/9781803230931

2. Submit your proof of purchase
3. That's it! We'll send your free PDF and other benefits to your email directly